HANDS-ON EXPERIENCE OF MICROBIAL WORLD

FIRST EDITION

HANDS-ON EXPERIENCE OF MICROBIAL WORLD

MICROBIOLOGY LABORATORY EXERCISES

KYU HONG CHO, PH.D.

Indiana State University

cognella

SAN DIEGO

Bassim Hamadeh, CEO and Publisher
Angela Schultz, Associate Acquisitions Editor
Craig Lincoln, Project Editor
Jeanine Rees, Production Editor
Jessica Delia, Graphic Design Associate
Laura Duncan, Licensing Coordinator
Natalie Piccotti, Director of Marketing
Kassie Graves, Senior Vice President, Editorial

Printed in the United States of America.

cognella® ACADEMIC PUBLISHING
3970 Sorrento Valley Blvd., Ste. 500, San Diego, CA 92121

BRIEF CONTENTS

DETAILED CONTENTS

EXERCISE 1

MICROORGANISMS IN OUR ENVIRONMENT

LEARNING OBJECTIVES

After this exercise, you should be able to

1. Define the meaning of ubiquity and diversity of microorganisms.
2. Explain the sources of bacteria and fungi in our environment.
3. Properly label and incubate agar plates.

Introduction

Even though you do not see microorganisms in most situations, they are around you everywhere. They are in the air, soil, and water, such as rivers and oceans, and on the desks, cell phones, and computers you are using now. Especially bacteria and bacterial spores, fungi and fungal spores, and viruses are everywhere. You cannot avoid microorganisms in your ordinary life. You are covered with microbes. They are on the skin, on the hair, inside your throat, and inside your intestines, especially the large intestine. Thus, microbes are ubiquitous, meaning they are everywhere in our environment. There are many kinds of microorganisms on the Earth. We do not know exactly how many kinds of microorganisms are on the Earth. According to a study, it is estimated that there are about one trillion species of microbes on the Earth, and most of them have yet to be discovered. In this exercise, you will experience the ubiquity and diversity of microorganisms by collecting samples in our environment.

The total weight of microorganisms on Earth is heavier than that of all other organic life combined, most of which are in the ocean. Microorganisms have significant roles in maintaining Earth's ecological balance. Two examples are given here among the many roles of microorganisms in this matter. We need oxygen to breathe, and probably some of you believe that most oxygen is produced by plants. As a matter of fact, some microorganisms can do photosynthesis like plants, and they produce more oxygen than plants. Some microorganisms can degrade organic matter quickly and efficiently. Animals and plants die all the time, but we do not see a pile of animals' dead bodies around us because microorganisms are degrading them fast. Humans use this microbial characteristic to clean water; microorganisms are used to clean sewage water.

Animals and plants are covered by microorganisms. We are not an exception. Microorganisms are on our body surfaces and in the respiratory and gastrointestinal tract. These microorganisms are called human microbiota. These microorganisms have important roles in maintaining our health. Our large intestine contains most of the microorganisms on our body. Around a third of feces are microbes. Some of these intestinal microbes digest food materials undigested by human enzymes and produce organic acids. These organic acids are absorbed in the large intestine and used as an energy source. Some of these intestinal microorganisms also produce vitamins such as vitamin K. Our immune system uses the human microbiota to train immune cells. Some of our immune cells interact with the microorganisms under the mucosal membranes and get information on the microorganisms. Sometimes, we suffer from infections caused by microbes. These disease-causing microbes are called microbial pathogens or pathogens. Not all microorganisms cause diseases. Only a few microorganisms can cause diseases in humans, animals, or plants. Most microorganisms in our environment do not cause human diseases, but many of them can spoil our food, clothes, and other materials when they grow. Knowing the ubiquity of microbes helps us properly respond to and solve the problems caused by them.

We use microorganisms to produce foods and chemicals that are used in our lives. Yeasts are used to make bread and beer. If beer is fermented longer, bacteria take up ethanol and produce acetic acid, the chemical substance of vinegar. Many microorganisms are also used to make diverse cheese. Traditionally, some industrial chemicals, such as ethanol and acetone, are also produced using microorganisms. Many enzymes, such as amylases, glucose isomerases, cellulases, and lipases, are produced using microorganisms. Amylase converts starch to glucose, and glucose isomerase converts glucose to fructose, which is sweeter than glucose. High fructose corn syrup, a sweetener in many beverages, is made from corn starch using amylases and glucose isomerases. These days, insulin that is used to treat diabetes mellitus is produced using microorganisms. When the human insulin gene is transferred to a microorganism by genetic engineering, the microorganism can produce human insulin.

All organisms on Earth are divided into prokaryotes or eukaryotes. Eukaryotic cells are much bigger and have more cellular structures than prokaryotic cells. Eukaryotic cells have subcellular organelles enclosed by lipid membranes such as the nucleus, mitochondria, lysosomes, Golgi apparatus, and chloroplasts. However, prokaryotic cells lack these membrane-enclosed organelles. Since the nucleus is the most prominent organelle, it is simply defined that the organisms with a nucleus are eukaryotes, and those without are prokaryotes. Even though prokaryotes are simpler than eukaryotic cells regarding their structures, they are similar in their chemical composition and chemical reactions.

In 1978, Carl Woese at the University of Illinois at Urbana-Champaign devised a system of classification based on the cellular organization of organisms into three domains. The three domains are bacteria, archaea, and eukarya. All bacteria and archaea are microorganisms. Eukarya contains four groups: protists, fungi, plants, and animals. Among them, protists and fungi are classified as microorganisms. Protists are divided into protozoa and algae, and fungi include yeasts, molds, and mushrooms.

Materials needed

- 1 TSA (tryptic soy agar) plate per student
- 1 PDA (potato dextrose agar) plate per student

Procedure (individual exercise)

Day 1

1. Draw a line in the middle on the bottom of the agar plates shown in **Figure 1.1**. You will make streaks of two samples on each plate. Write your initials and date on the bottom plate. When you label plates, label them on the bottom plates since the lids of the plates can be swapped.
2. Perform sampling using sterile cotton swabs and transfer the sample to agar plates. Gently move the cotton swab over the surface of the media shown in **Figure1.1**. Swab two places and streak gently on the agar plate to

demonstrate the ubiquity of microorganisms. Each sample will be streaked onto two plates (TSA and PDA). TSA plates are mainly for growing bacteria, and PDA plates are for fungi. You are allowed to do sampling outside of the lab (e.g., lab bench, floor, cell phone, laptop, keyboard, computer mouse, watch, finger, arm skin, nostril, hair, clothes, sink/faucet, doorknob or handle, coin, paper bill, wallet, paper towel, soil, bench, chair, stair, stair handrail, hallway floor, or toilet seat).

3. Write the sampling locations on the bottom plates shown in **Figure 1.1**.
4. Incubate plates at room temperature in the lab. Turn plates upside down to incubate. If you do not turn plates upside down, water vapor condensate can accumulate inside the plate's lid and drip onto the surface of the agar plate, ruining microorganism cultivation.

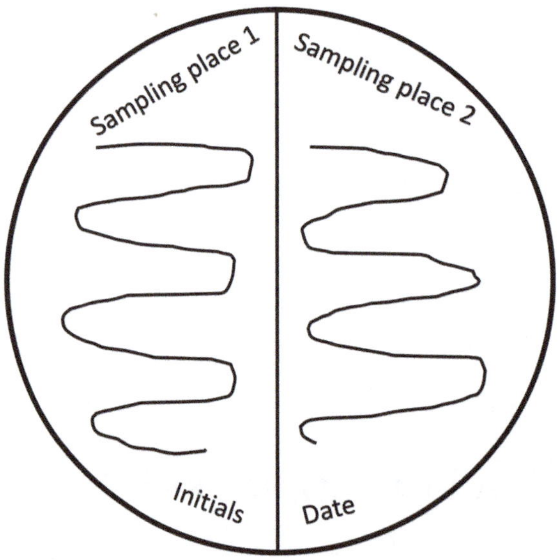

FIGURE 1.1 Labeling of plates and simple streaking of samples.

Day 2

1. Record your observation in **Table 1.1** on the bacterial growth on the plate. Make a drawing of what you observe, paying particular attention to the shape and the color of microorganisms.

TABLE 1.1 Draw and describe what you see on the plates.

TSA plate

Image	Description

PDA plate

Image	Description

Review questions

1. Explain the ubiquity and diversity of microorganisms.

2. Why is it important to be aware of the ubiquity of microorganisms during surgery or food processing?

3. Why are agar plates incubated upside down?

4. Why are agar plates labeled on the bottom plates?

EXERCISE 2

OBSERVATION OF MICROORGANISMS UNDER A BRIGHTFIELD MICROSCOPE

LEARNING OBJECTIVES

After this exercise, you should be able to

1. Explain the principles of light microscopy.
2. Identify the parts of a brightfield microscope and explain the function of each part.
3. Calculate the total magnification of light microscopy.
4. Know how to use a brightfield microscope.
5. Know when and why an oil immersion lens is used to see prokaryotic microorganisms.
6. Clean and properly store the microscope.

Introduction

In general, an individual microorganism is a tiny creature we cannot see with our naked eyes. We need a special instrument to see microbes. In this exercise, you will observe eukaryotic and prokaryotic microorganisms using a light microscope. The basic compound light microscope is called a brightfield microscope. The object or specimen being viewed is illuminated by transmitted light from beneath. Above the specimen, the light passes through the objective lens, which creates a magnified image in the microscope. The eyepieces or ocular lenses in turn project a magnified image into the eyes of the viewer (**Figure 2.1**).

To operate the microscope, three kinds of adjustments are required.

1. Selection of objective lens: Most light microscopes have several objectives, and they are generally 4X, 10X, 40X, and 100X (Table 2.1). The 4X and 10X objective lenses are useful for initial scanning and focusing. The 40X objective lens is the high dry lens and used to view most eukaryotic cells. The 100X objective lens is used to observe bacterial cells. The 100X objective lens is called the oil immersion lens because it needs oil to provide a clear image. The ocular lens has a magnification of 10X. The total magnification is equal to the objective magnification multiplied by the ocular magnification. If you use the 100X objective lens to see microorganisms, the total magnification is 1,000X (100X objective lens times 10X ocular lens). The objectives are located on a revolving nosepiece and are clicked into place using the black ring. Do not touch the objectives to rotate them. Instead, use the black ring.

2. Positioning of slide: There are two controls located below the stage on the right that move the slide. They work at right angles to each other and permit any part of the slide to be viewed.

3. Focusing: There are two focusing knobs. They adjust the distance between the stage and the objective lens. The larger knob is for the coarse focus. It should be used first with the 4X or 10X objective lens to bring the specimen approximately into focus. The smaller knob is for the fine focus. It is used with the 40X or 100X objective lens to obtain the sharpest possible image.

Total magnification is the ratio of the size of the image to the actual size of the object. A 1,000-fold magnification is about the maximum that can be obtained with a brightfield microscope. This limit is caused by the nature of visible light, so it is not possible to see any more detail in the object at higher magnifications than 1,000X. The sharpness of an image is called resolution. The resolution of a lens system is measured as the smallest distance between two points at which they appear clear and distinct from one another. Just how strong the resolving power is depends on the wavelength of light and on the amount of light that reaches the objective lens. Thus, proper illumination is essential for producing resolution. The function of the condenser is to focus and direct light from a source to the specimen. The condenser's position, the distance between the condenser and the stage, determines how accurately the light is focused on the end of the objective lens. The iris diaphragm controls the amount of light entering the condenser. As the magnifying power of a lens increases, the diameter decreases, and the lens admits less light. As a general rule, the amount of light being directed through the slide should be increased as the magnification is increased.

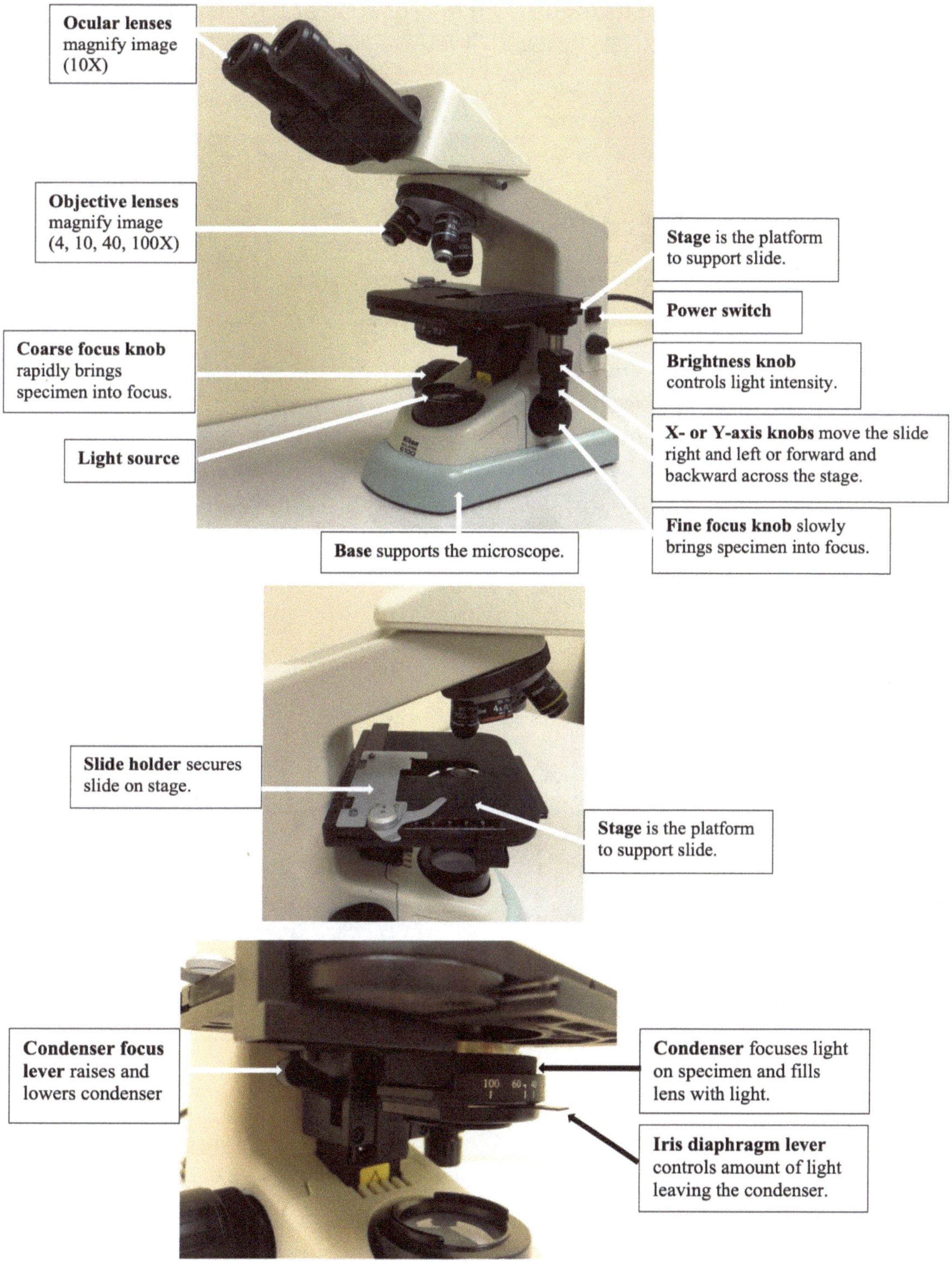

FIGURE 2.1 Parts of the microscope and their functions.

TABLE 2.1 Objective lenses of the light microscope and total magnification.

Objective lens magnification	Objective lens name	Usage	Total magnification
4X	Scanning lens	To scan samples	40X
10X	Low power lens	To initial search and focus on samples	100X
40X	High dry lens	To observe most eukaryotic cells	400X
100X	Oil immersion lens	To observe bacterial specimens	1,000X

The oil immersion objective is very small in diameter compared to the other objectives. Because its diameter is so small, much of the light from the specimen is not gathered and does not contribute to the formation of the image (**Figure 2.2**). This situation is improved by using immersion oil. This oil has a refractive index like that of glasses. Without oil, the light passes through the air between slide and objective. Air has a different refractive index from glass, so this causes some of the light to be deflected away and lost from the image. With the oil, no light is deflected.

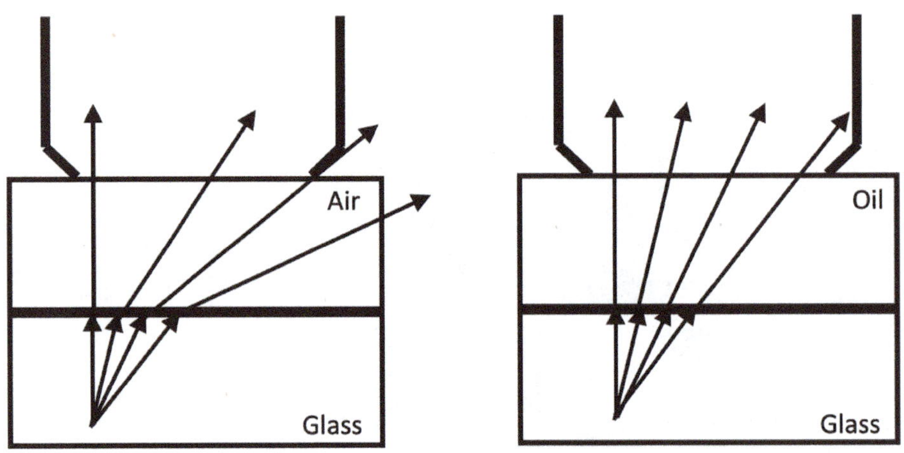

FIGURE 2.2 Passage of light from glass into air or oil. When a ray of light passes from glass (a denser material) into air (a less dense material), it is bent toward the surface between the two materials. However, it is not bent when light passes glass into oil (almost the same dense material). The example in the figure shows a light point from the light source, and in reality, there will be many light points from the light source.

Materials needed

- One microscope per student
- Microscope slides and cover slips, one per student
- Pond water or fish tank water sample, ~50–100 ml per class
- Bacterial demonstration slides
- Eukaryotic demonstration slides
- Lens paper
- Lens cleaner

Procedure 1: Setting up a microscope (individual exercise)

1. Place a microscope on the lab bench. You should use both hands when you handle microscopes. Use one hand to grasp the arm of the microscope and support it from the base with the other hand.
2. Clean both the ocular and the objective lenses with lens paper to obtain a clear image. Wipe each lens gently several times. Do not touch the lenses with your fingers.

Procedure 2: Examination of eukaryotic demonstration slides: low power and dry samples (individual exercise)

1. Plug in the microscope and turn it on.
2. Lower the mechanical stage and place the lowest power objective (4X) right above the stage.
3. Prepared eukaryotic demonstration slides will be provided. Mount a slide on the microscope by placing the slide in the slide holder on the stage. You should place the stain side up.

4. Adjust the condenser to a position close to the top of its travel.
5. Position the sample on the slide in the center over the condenser by moving the mechanical stage.
6. Make sure that the objective lens is positioned over the slide glass and raise the stage all the way to the top using the coarse focus knob.
7. Focus on the sample by slowly lowering the stage with the coarse focus knob to see a sharp image.
8. Once the sample is in focus under the objective, you do not need to use the coarse focus knob. Any further focusing should be done with the fine focus knob.*
9. Place the 10X or 40X objective in place by rotating the nosepiece. The lens of 10X objective or high dry objective (40X) (depending on the size of micro-organisms) is enough to see eukaryotic microorganisms.
10. Move the fine focus knob until the image is sharp. In any case, the objective lenses should not touch the slide glass.
11. To have a better image, adjust the iris diaphragm (open or close) and the condenser (lower or raise).
12. Observe two different eukaryotic demonstration slides and draw and describe what you see in **Table 2.2**.
13. Return the demonstration slides to the slide box.

* Most current microscopes are approximately in focus no matter what objectives are used. A microscope that has all the objectives approximately in focus at the same point of travel is said to be **parfocal**.

TABLE 2.2 Draw and describe what you see on the eukaryotic demonstration slides under the microscope.

Microorganism name	Image	Description

Procedure 3: Examination of water sample: low power and wet mount (individual exercise)

1. A pond or fish tank water sample will be provided in the lab. Place a drop of the water sample on a clean slide glass.
2. Place a coverslip on the water droplet. Since this is a liquid sample, you should use a cover slip to prevent the liquid from touching the lenses. To avoid trapping air during this step, place an edge of the coverslip at a 45-degree angle and gently lower the coverslip over the water drop.

3. Focus with the 4X objective. Follow the previous procedure to focus the microscope on the sample.
4. Observe the sample under the 10X or 40X objective lenses (depending on the size of microorganisms).
5. Adjust the iris diaphragm (open or close) and condenser (lower or raise) to see a sharp image.
6. Draw and describe your observation in **Table 2.3**.

TABLE 2.3 Draw and describe what you see in the water sample under the microscope.

Image	Description

Procedure 4: Examination of bacteria sample: high power and oil immersion (individual exercise)

1. Bacteria cells are so tiny, individual bacteria can be seen under only 100X objectives. Bacteria demonstration slides will be provided in the lab. Obtain a slide and mount it on the microscope.
2. Focus on the sample with the 4X objective.
3. Switch the objective to the 10X objective and focus on the sample.
4. Switch the objective to the 100X objective and focus on the sample. The 100X lens is called an oil immersion lens because it is used with immersion oil.

Immersion oil should never be used with other lower power objective lenses because these lenses cannot block the penetration of oil inside lenses. If oil enters inside 10X or 40X objective lens, this destroys the lens.

5. Rotate the nosepiece a little bit so that no objective is placed over the stage.
6. Add one or two drops of immersion oil right above the bacterial cells.
7. Position the 100X objective into place.
8. Adjust the iris diaphragm all the way open and the condenser all the way up to have enough light and contrast.
9. Focus on the sample using the fine focus knob until you have a sharp image. Do NOT use the coarse focus knob.
10. If you get lost, go back to the 4X or 10X objective and refocus, then switch back to the 100X objective. Once there is oil on the slide, make sure the 40X objective does not touch oil.
11. Draw and describe your observation in **Table 2.4**.
12. Wipe the oil off the demonstration slides and return them to the slide box.
13. Clean the 100X oil immersion objective lens with lens paper and lens cleaner solution provided in the lab. Repeat the cleaning with a fresh piece of lens paper until the lens paper does not become oily.

TABLE 2.4 Draw and describe what you see on the bacteria sample slide under the microscope.

Label of the slide ______________________________

Image	Description

Procedure 5: Cleaning and storing a microscope

1. Place the 4X objective in the view position.
2. Clean up all the lenses with lens paper.
3. Clean up any oil on the microscope using a paper towel.
4. Turn off the power.
5. Unplug the electrical cord and wrap it on itself and hang it from the ocular.
6. Store in a cabinet.

Review questions

1. If you choose 40X objective lens, what is the total magnification?

2. Explain parfocal.

3. What is the role of the condenser?

4. Which one needs more light for a clear image—100X objective lens or 10X objective lens?

5. Which objective lens should be used to observe bacteria?

6. To observe bacteria under a light microscope, immersion oil is used. What is the purpose of the oil?

EXERCISE 3

BACTERIAL COLONY MORPHOLOGY

LEARNING OBJECTIVES

After this exercise, you should be able to

1. Define medium, inoculum, and culture.
2. Define solid medium and liquid medium.
3. Know the definition of a microbial colony.
4. Know how to describe colony morphologies of microorganisms.

Introduction

To grow microbes, the proper nutrients and growth conditions are required. The foods used to grow microbes are called media (singular: medium). Media contain all the nutrients for microbes to grow. Before the introduction (or inoculation) of a new microorganism you want to grow, culture media must not have any other microbes. This state is called sterile. An inoculating loop is commonly used to introduce microbes to media. First, all the microorganisms on the inoculating loop are killed by flame. After being cooled down, the inoculating loop touches a source of microorganisms and introduces them to media. The microorganism introduced to sterile media is called inoculum. If you incubate the inoculated media in the right condition, microbes will multiply, so there will be many microbes in the media. The media with microbes are called cultures or microbial cultures.

Media are broadly divided into liquid or solid media. Liquid media are commonly called broth, and solid media are called agar plates. Generally, solid media are used to grow cells on the surface of the media. Agar is used to make solid media. Agar is a complex polysaccharide complex originating from seaweeds. The physical property of agar is like jelly or gelatin. However, unlike jelly, agar is not degraded by microbes. Agar is sold as a powder form. Agar melts at 100°C, so when a liquid media with agar powder is autoclaved at 121°C, agar becomes liquid. Agar solidifies at ~40°C. When the autoclaved media are cooled down and reach room temperature, the media become solid.

When bacteria are spread on agar plates and incubated for enough time, they form colonies, which are visible masses consisting of millions of cells through multiple divisions. Each colony arises from an individual cell (in the case of coccus or bacillus), or a group of cells (a string in the case of *Streptococcus* or *Streptobacillus* or a clump in the case of *Staphylococcus*). A colony is considered a pure culture because it consists of the same cells. Each bacterial species generally has a unique colony appearance or morphology in terms of size, shape, color, and the like (**Figure 3.1**), so the observation of colony morphology is often the first step in the identification process of bacterial species.

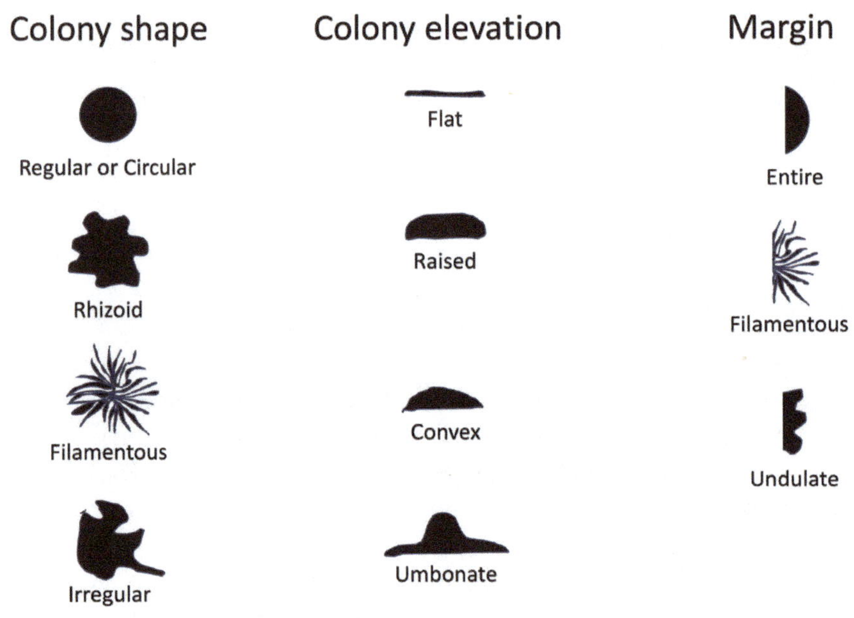

FIGURE 3.1 Bacterial colony morphology on solid media.

Materials needed

- TSA plates streaked with *E. coli*, *Bacillus subtilis*, *Mycobacterium smegmatis*, *Pseudomonas aeruginosa*, *Proteus mirabilis*, *Serratia marcescens*, *Micrococcus luteus*, *Staphylococcus aureus*, *Staphylococcus epidermidis*, *Streptococcus pyogenes*, *Streptococcus agalactiae*, *Streptococcus mutans*, *Streptococcus salivarius*

(3 plates of each microorganism per class. Prepare streak plates and incubate them at 37°C. Once visible colonies are shown, store them at 4°C; plates should be stored upside down.)

Procedure: Observation of bacterial colony morphology on plates (individual exercise)

1. Demonstration plates will be provided. Pick three plates from the demonstration plates and observe them. Identify their morphologies and fill in the blanks in **Table 3.1**. Use the terminology in **Figure 3.1**.

TABLE 3.1 Write the name of three microorganisms you picked and describe their colony morphologies.

Name of microorganism			
Size[1]			
Shape[2]			
Margin[3]			
Color[4]			
Elevation[5]			
Surface texture[6]			
Opacity[7]			

1: in millimeters (mm)
2: regular or circular, rhizoid, filamentous, irregular
3: entire, filamentous, undulate
4: red, purple, pink, yellow, green, cream, white, etc.
5: flat, raised, convex, umbonate
6: smooth, glistening, rough, wrinkled, dry
7: transparent, translucent, opaque

Review questions

1. Explain what inoculum is.

2. What characteristics of agar are used to make a solid medium?

3. What is a colony? Does a colony always arise from a single cell? Why are colonies considered a pure culture?

Credit

Fig. 3.1: Ewen, "Bacterial colony morphology of microorganisms on solid media," https://commons.wikimedia.org/wiki/File:Bacterial_colony_morphology.png, 2007.

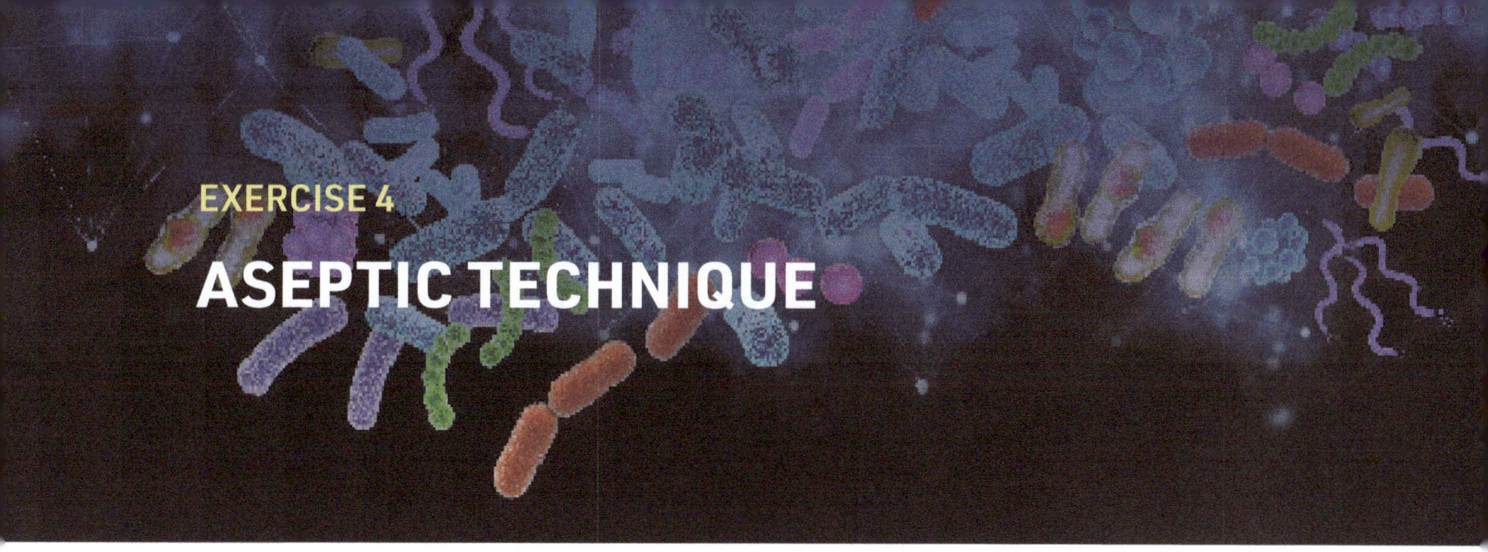

EXERCISE 4

ASEPTIC TECHNIQUE

LEARNING OBJECTIVES

After this exercise, you should be able to

1. Know what aseptic techniques are.
2. Explain the importance of aseptic techniques.
3. Describe how to achieve an aseptic environment in research labs and hospital settings.

Introduction

Microorganisms are everywhere, including in the air. To grow pure culture (a culture consisting of only one type or strain of microorganism) or prevent introducing other microorganisms when you transfer microorganisms from one place to another, aseptic techniques are required. The term *aseptic* means "without infection." Aseptic techniques are procedures or methods used to prevent contamination by microorganisms. This concept is also required when healthcare professionals treat patients. During surgery or injections, healthcare professionals must avoid microbial infections. To do this, the surgical instruments and needles should be sterilized and carefully handled, and the surgical location or the injection part of a patient's body should be cleaned properly using chemicals to reduce the number of microorganisms significantly. These are all considered aseptic techniques.

In this lab, you will mainly use a flame to achieve an aseptic environment. The inoculation loops and the top surface of glass tubes will be sterilized using a flame.

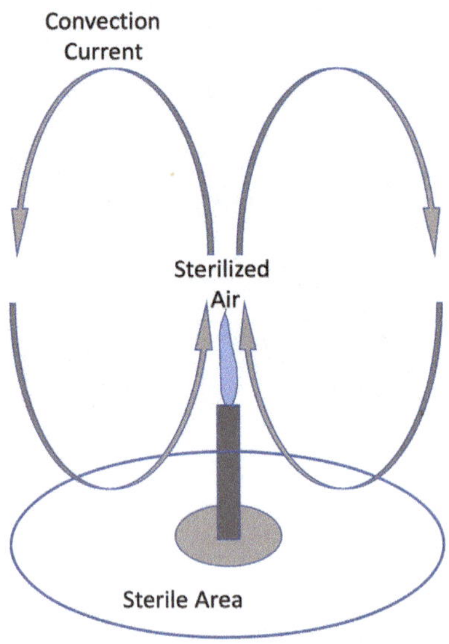

FIGURE 4.1 Convection around the flame of a Bunsen burner.

Also, you should perform transferring microorganisms or medium near flame (but not too close to it) because flame creates an aseptic environment through air convection currents (**Figure 4.1**).

Materials needed

- 3 capped tubes containing 5 ml of sterile tryptic soy broth per student
- 1 *E. coli* broth tube per table

Procedure: Inoculation into broth tubes (individual exercise)

Day 1 (the instructor will give a demonstration)

1. Label the tubes of broth "tube 1," "tube 2," or "tube 3," your initials, section number, and today's date with a Sharpie pen, and place both tubes in a test tube rack.
2. Light your Bunsen burner.

3. Obtain a loop full of broth from tube 1 and transfer it to tube 2 based on the following steps (refer to **Figure 4.2**).
    a. Hold the inoculation loop between the thumb and the index finger with the wire loop pointing away from you (**Figure 4.2.a**).
    b. Flame the entire wire loop until it is red hot (**Figure 4.2.b**).
    c. Then, remove the inoculation loop from the flame and allow it to cool. This procedure sterilizes the loop (**Figure 4.2.c**).
    d. Using the little and ring fingers, remove the cap from tube 1. Do not set the cap down on the table during the procedure (**Figure 4.2.d & 4.2.e**).
    e. Briefly flame the lip of the tube. This procedure sterilizes the lip of the tube (**Figure 4.2.f**).
    f. Insert the cooled loop directly into the broth (**Figure 4.2.g**).
    g. Withdraw the loop.
    h. Reflame the lip of the tube (**Figure 4.2.h**).
    i. Replace the cap (**Figure 4.2.i**).
    j. Set the tube back in the test tube rack.
    k. Using the little and ring fingers, remove the cap from tube 2.
    l. Briefly flame the lip of the tube.
    m. Insert a loop full of broth directly into tube 2.
    n. Shake the loop several times.
    o. Withdraw the loop.
    p. Reflame the lip of tube 2.
    q. Replace the cap. Set the tube back in the test tube rack.
    r. Flame the entire wire loop until it is red hot.
    s. Remove the inoculation loop from the flame and allow it to cool.
4. Obtain a loop full of *E. coli* broth and transfer it to tube 3.
    a. Hold the inoculation loop between the thumb and the index finger with the wire loop pointing away from you.
    b. Flame the entire wire loop until it is red hot.
    c. Then, remove the inoculation loop from the flame and allow it to cool.
    d. Using the little and ring fingers, remove the cap from the *E. coli* broth tube.
    e. Briefly flame the lip of the tube.
    f. Insert the cooled loop directly into the *E. coli* broth.
    g. Withdraw the loop.

h. Reflame the lip of the tube.
i. Replace the cap. Set the *E. coli* tube back in the test tube rack.
j. Using the little and ring fingers, remove the cap from tube 3.
k. Briefly flame the lip of the tube.
l. Insert the loop with a loop full of *E. coli* broth directly into tube 3. Shake the loop several times.
m. Withdraw the loop.
n. Reflame the lip of tube 3.
o. Replace the cap. Set the tube back in the test tube rack.
p. Flame the entire wire loop until it is red hot and allow it to cool.

5. Place tubes 1, 2, and 3 in a 37°C shaking incubator (shaking speed: 200–300 rpm).
6. The instructor will remove the culture from the shaking incubator and store at 4°C in 24–48 hours.

Day 2

1. Observe the three tubes of broth for turbidity and record the results in **Table 4.1**.

TABLE 4.1 Growth of microorganisms in tubes tested.

	Tube 1	**Tube 2**	**Tube 3**
Microbial growth*			

- +++: high growth
- ++: medium growth
- +: low growth
- -: no growth

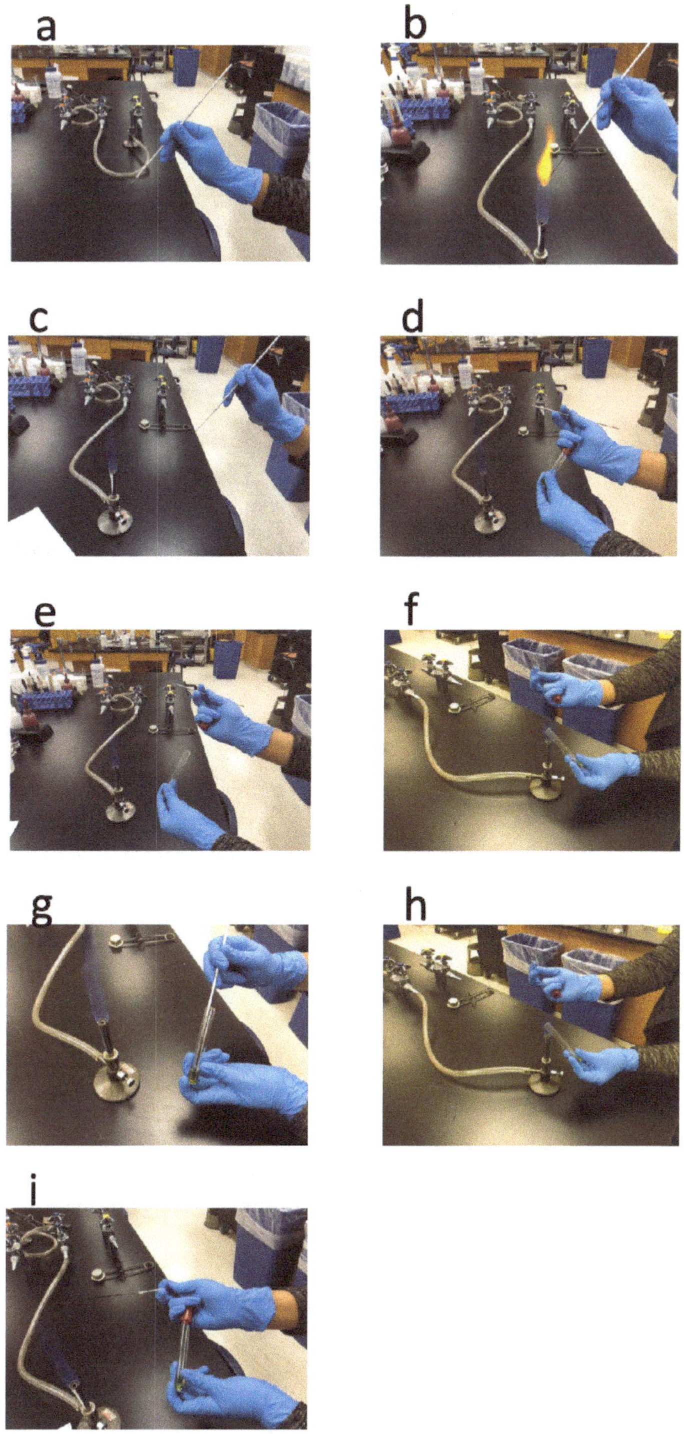

FIGURE 4.2 Aseptic culture transfer technique.

Review questions

1. What is the definition of aseptic techniques?

2. How are aseptic techniques applied in hospital settings?

3. How does a Bunsen burner flame create an aseptic environment?

4. Did you have the expected results in Table 4.1? If not, what procedures can be improved to achieve proper aseptic techniques?

EXERCISE 5

PURE CULTURE TECHNIQUE

The Streak Plate Method

LEARNING OBJECTIVES

After this exercise, you should be able to:

1. Define pure culture.
2. Explain the purpose of the streak plate method.
3. Perform the streak plate method and isolate single colonies.

Introduction

Generally, environmental and biological samples contain many kinds of microbes. To study a particular microorganism or identify a pathogen in an infected tissue, the microbe should be separated from other environmental microbes and normal microbiota. Once a microbe is isolated, a pure culture that contains one species or strain in a culture medium can be grown. To isolate a single microbe from other microbes, several methods can be used. However, the most popular and easiest method is the streak plate method. In this exercise, you will learn how to perform the streak plate method.

Materials needed

- Mixed broth culture of *Escherichia coli*, *Serratia marcescens*, and *Staphylococcus aureus* per lab bench: grow them separately in a 37°C shaking incubator overnight and mix them with an equal ratio
- Two TSA plates per student

Procedure: Single colony isolation by streaking on an agar plate (individual exercise)

Day 1 (the instructor will give a demonstration)

1. Get two TSA agar plates from the instructor and label them with your name and lab section number on the back of the plate.
2. Place the plates upside down on the lab bench.
3. Perform streak plating **twice**. Refer to **Figure 5.1**.
    a. Flame your inoculating loop and cool down the loop by touching the agar surface of the agar plate.
    b. Get a loop full of bacteria from the mixed culture.
    c. Streak the loop on a corner of the plate (first streak).
    d. Flame the loop and cool down by touching down the uninoculated agar surface.
    e. Perform the second streak. Pass it once through the already-streaked region of the plate and streak another small portion of the plate.
    f. Flame the loop again and cool down by touching down the uninoculated agar surface.
    g. Make the third streak. Pass it once through the second-streaked region of the plate and streak another small portion of the plate.
    h. Flame the loop again and cool down by touching down the uninoculated agar surface.
    i. Make the fourth streak. Pass it once through the third-streaked region of the plate and streak another small portion of the plate.
4. Incubate the plate at 37°C for 36–48 hrs. Place the plate upside down. If you incubate agar plates top side up, moisture droplets will appear inside the lid and eventually drop onto the agar. This will cause cells to spread on the agar instead of forming colonies.

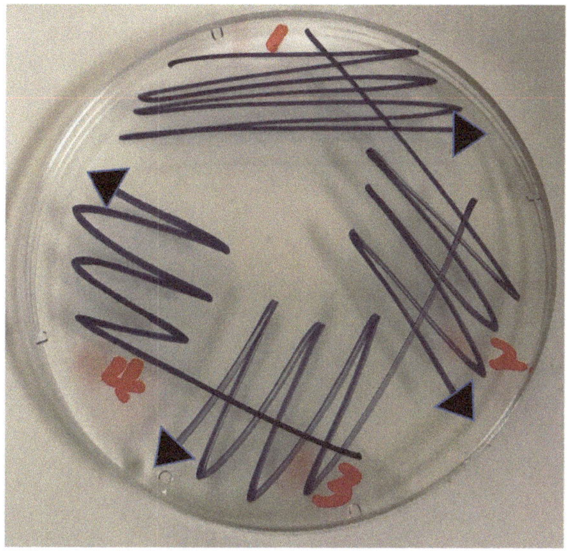

FIGURE 5.1 The streak plate method for isolating microbes. Arrow heads indicate the direction of streaking. The first streak is made from the original bacterial culture. The inoculating loop is sterilized following each streak. In the second, third, and fourth streaks, the loop picks up bacteria from the previous streaks, diluting the number of cells each time.

Day 2

1. Record your observation on the bacterial growth on the plate in **Table 5.1**. Make a drawing of what you observe, paying particular attention to the colony morphologies, relative sizes, and colors of colonies. A single colony usually appears in the third or fourth streak region, depending on the initial bacterial density and your technique.

TABLE 5.1 Draw and describe what you observe on the plate.

Plate	Description

Review questions

1. What are pure cultures?

2. What is the most popular way to obtain single colonies on a plate and how is it done?

3. Did you get single colonies on your plate? If not, what steps in your streak plate method do you think should be improved?

EXERCISE 6

SIMPLE STAINS AND GRAM STAIN

LEARNING OBJECTIVES

After this exercise, you should be able to

1. Explain why microorganisms need to be stained when they are observed under a microscope.
2. Distinguish simple staining and differential staining.
3. Describe the main advantage of differential staining over simple staining.
4. Prepare bacterial smear.
5. Evaluate and improve your simple staining and Gram staining techniques.
6. Illustrate the cell wall structures of Gram-positive and Gram-negative bacteria.

Introduction

Most microorganisms do not have color constituents (pigments) inside cells, so it is not easy to see them even under a microscope. Thus, several staining methods have been developed to see microorganisms better under a brightfield microscope. Simple staining methods employ just one chemical that stains almost all microorganisms, so they are simple and convenient. Simple stains can be used to figure out microorganisms' size, shape, and arrangement (**Figure 6.1**).

Most bacteria are around 1 micrometer in diameter and 2 to 8 micrometers in length. These tiny creatures have several distinct shapes. Many bacteria are rod-shaped and called bacillus (plural: bacilli). Spherical or ball-shaped bacteria are called coccus (plural: cocci). Twisted-shaped bacteria are called spirals. There are three types of spirals: vibrio, spirillum, and spirochete. They are classified based on the degree of twist. Vibrios have a curved rod shape. Spirilla have a helical shape like a corkscrew and are rigid. Spirochetes have a long helical structure and are more flexible than spirilla. Some bacteria have several shapes, not just one, and these bacteria are called pleomorphic bacteria.

Some bacteria are not separated after cell division and form a string or cluster of cells. If two cocci are attached, they are called diplococci. If cocci form a string or chain, they are called streptococci. Some cocci form a cluster because the plane of

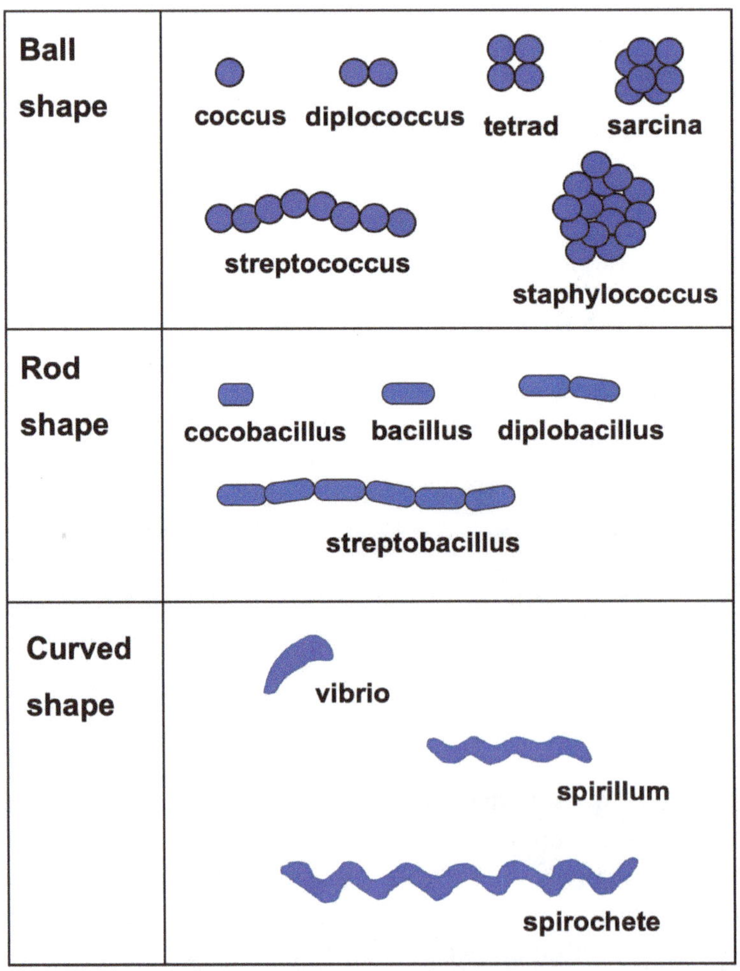

FIGURE 6.1 Bacterial shapes and arrangement.

division alternates between generations, such as north-south, then east-west, then they become staphylococci. If two bacilli stick together, they are called diplobacilli. If bacillus cells form a string or chain, they become streptobacilli. Staphylobacilli exist but are not common.

Some staining methods can differentiate microorganisms' types or show special cellular structures such as endospores or flagella. These staining methods are called differential staining. Examples of differential stains are Gram stain, acid-fast stain, flagella stain, and endospore stain. In this exercise, you will learn simple stain and Gram stain. Acid-fast stain and endospore stain will be covered in the next exercise.

Before staining, bacterial cells need to be fixed on a slide glass to prevent being washed out during staining processes. This fixation is achieved by drying bacteria samples on a slide glass followed by brief heating, known as heat fixation.

Simple staining

Simple staining methods use one chemical to stain microorganisms. The outer surface of most bacteria is negatively charged, so it binds well to positively charged dyes (basic dyes) such as crystal violet, safranin, and methylene blue. In this exercise, methylene blue will be used for simple staining.

Gram staining

Gram staining was invented by the Danish microbiologist Hans Christian Gram in 1882. Since then, this staining method has been widely used to differentiate two large groups of bacteria: Gram-negative and Gram-positive. It is typically the first test performed to identify a bacterial species.

Bacterial cell walls have two primary functions. They maintain cell shapes and protect cells from changes in osmotic pressure. Most bacteria live in a liquid environment or are exposed to a liquid environment at some point in their lives. Most water, such as river water and rainwater, has fewer salts than the cytoplasmic fluid inside bacterial cells. This environment is hypotonic, meaning fewer salts. When bacteria are in hypotonic liquid, water enters the cells by osmotic pressure. If there are no cell walls, then bacteria will be lysed. Bacterial cell walls are mainly made of peptidoglycan, which eukaryotic cells do not synthesize. If the synthesis of peptidoglycan is inhibited, bacteria die due to osmotic lysis. Almost half of the antibiotics currently used to cure bacterial infections inhibit bacterial cell wall synthesis.

Gram staining shows different colors of tested bacteria depending on bacterial cell wall structures. Gram-positive bacteria have a thick cell wall made of a carbohydrate-peptide complex called peptidoglycan. Gram-negative bacteria have a thin layer of peptidoglycan, and this is surrounded by another layer of an outer membrane comprising mainly phospholipids and lipopolysaccharides (LPS) (**Figure 6.2**).

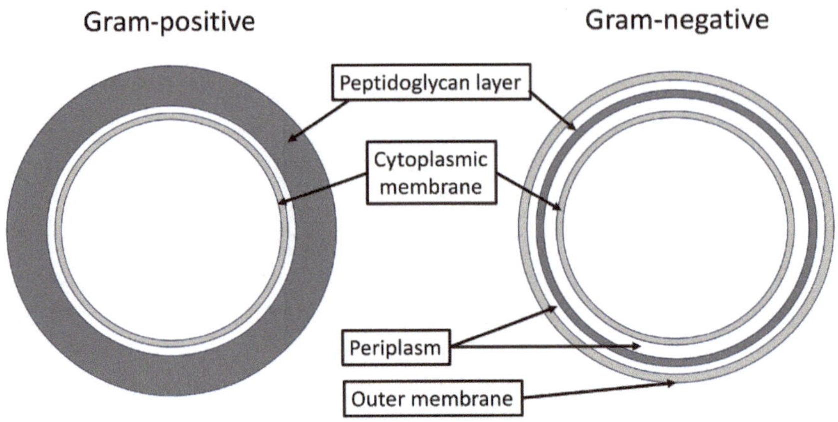

FIGURE 6.2 Cell wall structure of Gram (+) and Gram (−) bacteria.

For Gram stain, after the heat-fixation, cells are stained with crystal violet (a purple dye) (**Figure 6.3**). After cells are washed with water, the second chemical, iodine, is applied to the cells. Iodine is a mordant that makes a complex with crystal violet. This crystal violet–iodine complex does not come out of cells easily when the cells are washed with alcohol. Alcohol dehydrates the peptidoglycan of Gram-positive cells to make it more impermeable to the crystal violet–iodine complex. However, the effect on Gram-negative cells is different. Alcohol dissolves the outer membrane of Gram-negative cells and leaves small holes in the thin peptidoglycan layer through which the crystal violet–iodine complex diffuses. After this step, Gram-positive cells are purple, but Gram-negative cells don't have any color. To provide a color to Gram-negative cells, safranin, the counterstain, is applied. Safranin is a pink dye and stains all cells with pink. However, the color of Gram-positive cells is not changed since the pink color is obscured by the purple color.

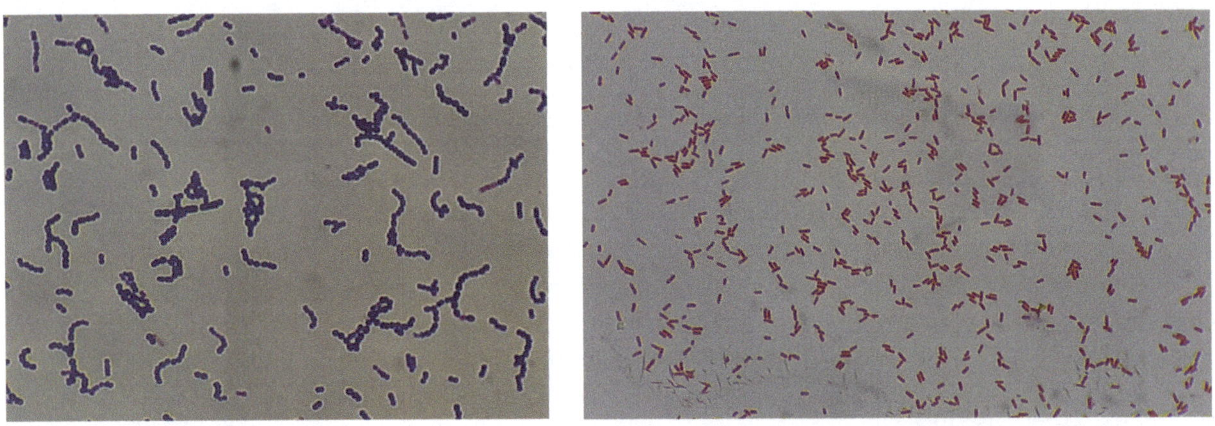

FIGURE 6.3 Examples of Gram staining results. A. Gram-positive streptococcus (purple). B. Gram-negative bacillus (red).

Materials needed

- TSB liquid cultures of bacteria: *Staphylococcus aureus*, *Escherichia coli*, *Bacillus subtilis*, *Streptococcus pyogenes* (1 tube of each bacterium per lab bench)
- TSA agar plate cultures of bacteria: *Staphylococcus aureus*, *Escherichia coli*, *Bacillus subtilis*, *Streptococcus pyogenes* (1 agar plate of each bacterium per lab bench)
- Microscope slides (2 per student)
- Methylene blue solution
- Crystal-violet solution
- Gram-iodine solution
- Safranine solution
- Bottles containing 95% ethanol or acetone (decolorizer)
- Wash bottles
- A tube of sterile water (10 ml) per student

Procedure 1: Smear preparation (individual exercise)

1. Each student will need to make four smears: two for simple stains (one from a broth culture and one from an agar plate) and two for Gram stains (one from a broth culture and one from an agar plate). Pick the same microorganism of broth culture and agar plate (e.g., *E. coli* broth culture and agar plate, and *S. pyogenes* broth culture and agar plate).
2. Label the initials of the microorganisms you picked near the edge of microscope slides (**Figure 6.4**).

Smear of cells from a broth culture

3. Mix bacterial suspension gently by tapping the bottom of the tube as demonstrated by the instructor and aseptically remove a loop full of a culture. Flame the lip of the tube and replace the cap.
4. Transfer the loop full of culture to the left side of the slide you have labeled.
5. Spread the culture over the area about the size of a dime (**Figure 6.4**).

6. Completely air dry the slide. If this takes more than fiveminutes, you probably have not spread the culture out properly.

Smear of cells from an agar plate

7. Place one loop full of sterile water on the right side of the same slide.
8. With the inoculating loop, take a little bit of bacterial cells from an agar plate and add the bacterial cells to the drop of water as demonstrated by the instructor.
9. Spread the culture over the area about the size of a dime (**Figure 6.4**).
10. Completely air dry the slide.

Heat fixation

11. Heat-fix the bacterial specimen by passing the slide slowly over the flame twice as demonstrated by your instructor. This heat fixation kills bacteria and makes bacteria adhere to the slide, so they are not rinsed off the slide during staining.

Preparation of another set

12. Perform the smear preparation with the other bacterium (one broth and one agar plate with the same microorganism).

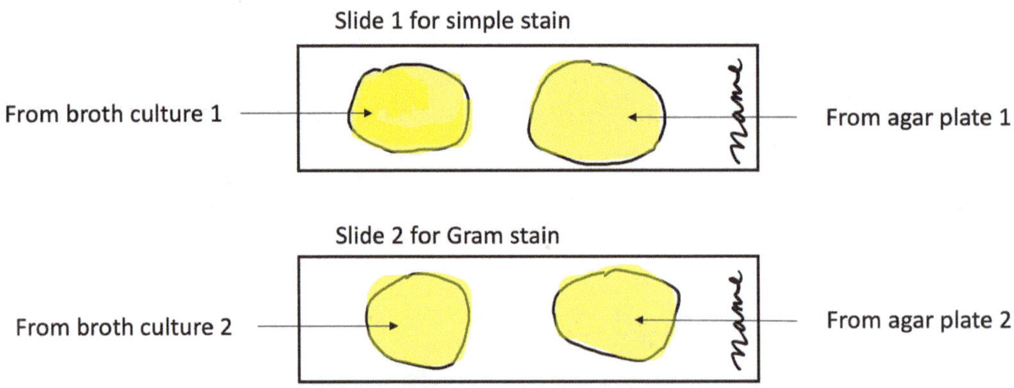

FIGURE 6.4 Slide preparation for simple and Gram staining.

Procedure 2: Simple stain (individual exercise)

1. Pick one smear slide you prepared.
2. Attach a clothespin to the end of the slide where microorganism's initials are located and place it on a staining rack over the sink.
3. Cover the smear with the **methylene blue** solution and wait two to three minutes.
4. Tilt the slide over a sink and wash excess stain off the slide with water in a wash bottle.
5. Carefully remove the remaining water drops by blotting the slide between paper towels.
6. Observe the bacteria on the slide using the oil immersion lens (100X).
7. Record your observation in **Table 6.1**. Make a drawing of what you observe and describe the shapes (morphology), relative sizes, and colors of the stained cells.

TABLE 6.1 After the simple staining, draw and describe what you see under the microscope (pick the better one out of the two).

Name of the bacterium ________________________________

Image	Description

Procedure 3: Gram stain (individual exercise)

1. Attach a clothespin to the end of the slide where microorganism's initial is located and place it on a staining rack over the sink.
2. Cover the smear with the **crystal violet** solution and wait one to two minutes.
3. Tilt the slide over the sink and rinse gently with water in a wash bottle.
4. Cover the smear with Gram's **iodine** for two minutes.
5. Wash gently with water.
6. Drip **95% ethanol** over the tilted slide until the drops run colorless. This decolorization is the most critical step in the Gram stain. Do not over-decolorize your smear. You should stop washing immediately when the wash drops become colorless. Decolorization will take no more than 15–20 seconds.
7. Cover the smear with the counterstain **safranin** for two minutes.
8. Rinse the slide with water.
9. Carefully remove the remaining water drops by blotting the slide between paper towels.
10. Observe the bacteria on the slide using the oil immersion lens (100X).
11. Record your observation in **Table 6.2**. Make a drawing of what you observe and describe the shapes (morphology), relative sizes, and colors of the stained cells.

TABLE 6.2 After Gram staining, draw and describe what you see under the microscope (pick the better one out of the two).

Name of the bacterium ______________________________

Image	Description

Review questions

1. Why do microorganisms need to be stained when they are observed under a microscope?

2. What is the advantage of doing a differential staining compared to a simple staining?

3. How does the simple staining work?

4. How does the Gram staining work?

5. What structure of a microorganism determines the color after Gram staining?

Credits

Fig. 6.3A: Copyright © by Ajay Kumar Chaurasiya (CC BY-SA 4.0) at https://commons.wikimedia.org/wiki/File:Viridans_strptococci_in_Gram_stain.jpg.
Fig. 6.3B: Copyright © by Ajay Kumar Chaurasiya (CC BY-SA 4.0) at https://commons.wikimedia.org/wiki/File:Gram_Negative_bacilli_of_Klebsiella_in_Gram_stained_smear_of_culture.jpg.

EXERCISE 7

ACID-FAST STAINING AND ENDOSPORE STAINING

LEARNING OBJECTIVES

After this exercise, you should be able to

1. Describe why methods other than Gram staining are used to stain *Mycobacterium* spp. and endospores.
2. Describe the structure of *Mycobacterium* cell walls.
3. Explain which microorganisms produce endospores.
4. Apply acid-fast staining.
5. Apply endospore staining.

Introduction

Some bacteria have atypical cell walls, so they are neither Gram-positives nor Gram-negatives. One bacterium containing an atypical cell wall is *Mycobacterium* species. Some *Mycobacterium* species cause serious human diseases. For example, *M. tuberculosis* causes tuberculosis, and *M. leprae* causes leprosy. *Mycobacterium* spp. have acid-fast cell walls. They have high concentrations (60%) of a waxy lipid called mycolic acid on top of a thin peptidoglycan layer. Mycolic acids are attached to the peptidoglycan layer, and this attachment is mediated by sugar polymers called arabinogalactan. The waxy lipid prevents the uptake of the Gram staining dyes. To stain *Mycobacterium*, a dye Carbol-Fuchsin is applied with heat that enhances penetration of the stain. Carbol-Fuchsin penetrates the cell wall, binds to the cytoplasm, and resists removal by washing with even acid-alcohol (**Figure 7.1**). That is why the cell walls are called acid-fast, meaning resistant to washing with acid.

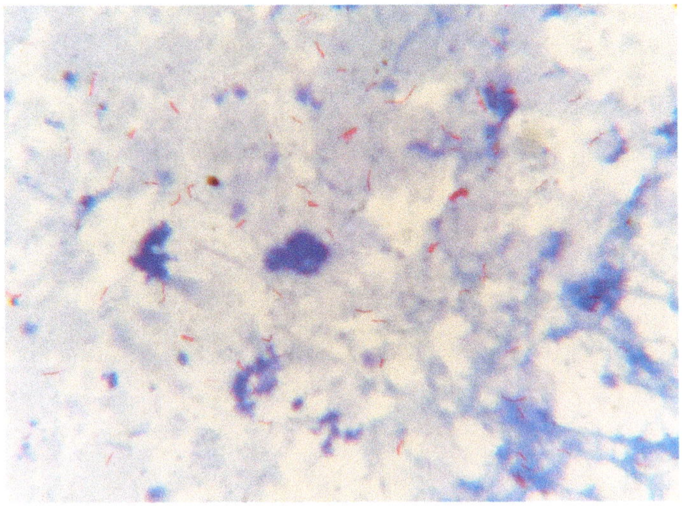

FIGURE 7.1 Acid-fast staining showing *Mycobacterium tuberculosis* (red) in a patient's sputum smear.

Endospores are resting structures formed inside of bacteria. Not all bacteria produce endospores; only some bacteria such as *Bacillus* and *Clostridium* produce endospores. Since they are common soil bacteria, their endospores are everywhere like fungal spores. The process of endospore formation is called sporulation, and the return of an endospore to its vegetative state is called germination.

Endospores resist staining with basic stains, so a nonbasic stain, malachite green, is used to stain endospores. Vegetative cells are stained with safranin (**Figure 7.2**).

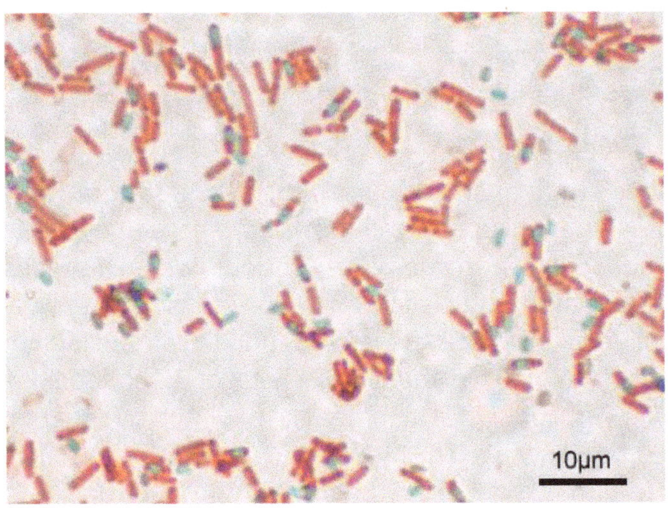

FIGURE 7.2 An endospore stain of *Bacillus subtilis*. Spores appear green due to the applied heat and application of malachite green stain. The *Bacillus subtilis* cells appear a reddish color due to the safranin counterstain.

Materials needed

- 1 broth culture of *Mycobacterium smegmatis* per lab bench
- 1 broth culture of *Bacillus subtilis* per lab bench (this culture should be grown for several days to make cells produce endospores)
- Carbol-Fuchsin
- methylene blue
- acid-alcohol
- malachite green

Procedure 1: Smear preparation (individual exercise)

1. Each student will need to make four smears: two for acid-fast staining (*Mycobacterium smegmatis* and *Bacillus subtilis*) on a slide glass and two for endospore staining (*Mycobacterium smegmatis* and *Bacillus subtilis*) on the other slide glass.
2. Label microscope slides near the edge with the initials of micro-organisms.
3. Gently swirl the bacterial suspension and aseptically remove a loop full of a culture. Flame the lip of the culture tube and replace the cap.
4. Transfer the loop full of culture to a slide you have labeled.
5. Spread the culture on the slide over the area of about the size of a dime.
6. Completely air dry the slide.
7. Heat-fix the bacteria specimen by passing the slide slowly over the flame twice as demonstrated by your instructor.
8. Perform the smear preparation with the other culture.

Procedure 2: Acid-fast staining (individual exercise)

1. Attach a clothespin to a slide with smears (to the end with microbes' initials) and place it on a staining rack over the sink.
2. Cover the smear with the **Carbol-Fuchsin** solution and leave on for five to seven minutes.
3. Tilt the slide over the sink and rinse gently with water. Not all the stain will be removed by water in this step.
4. Drip **acid-alcohol** over the tilted slide until the drops run colorless.
5. Rinse the slide with water.
6. Cover the smear with the counterstain **methylene blue** for two to three minutes.
7. Rinse the slide with water.
8. Carefully remove the remaining water drops by blotting the slide between paper towels.
9. Observe the bacteria on the slide glass using the oil immersion lens (100X).
10. Record your observation in **Table 7.1**. Make a drawing showing what you observe, paying particular attention to the shapes (morphology), relative sizes, and colors of the stained cells.

TABLE 7.1 Draw and describe what you see under the microscope.

Name of the bacterium ______________________________

Image	Description

Name of the bacterium ______________________________

Image	Description

Procedure 3: Endospore staining (Schaeffer-Fulton staining) (individual exercise)

1. Attach a clothespin to one end of a slide with smears (to the end with microbes' initials) and place it on a staining rack over the sink.
2. Cover the smear with the **malachite green** solution and leave on for more than 10 minutes. Malachite green stains endospores.
3. Tilt the slide over the sink and rinse gently with water.

4. Cover the smear with the counterstain **safranin** for two to three minutes. Safranin stains vegetative cells.
5. Rinse the slide with water.
6. Carefully remove the remaining water drops by blotting the slide between paper towels.
7. Observe the bacteria on the slide using the oil immersion lens (100X).
8. Record your observation in **Table 7.2**. Make a drawing of what you observe, paying particular attention to the shapes (morphology), relative sizes, and colors of the stained cells.

TABLE 7.2 Draw and describe what you observe under the microscope.

Name of the bacterium ____________________

Image	Description

Name of the bacterium ____________________

Image	Description

Review questions

1. Why does Gram staining not work for *Mycobacterium* spp.?

2. What staining method is used to stain *Mycobacterium* spp.?

3. Explain why endospores are everywhere.

4. Why does Gram staining not work for endospores?

Credits

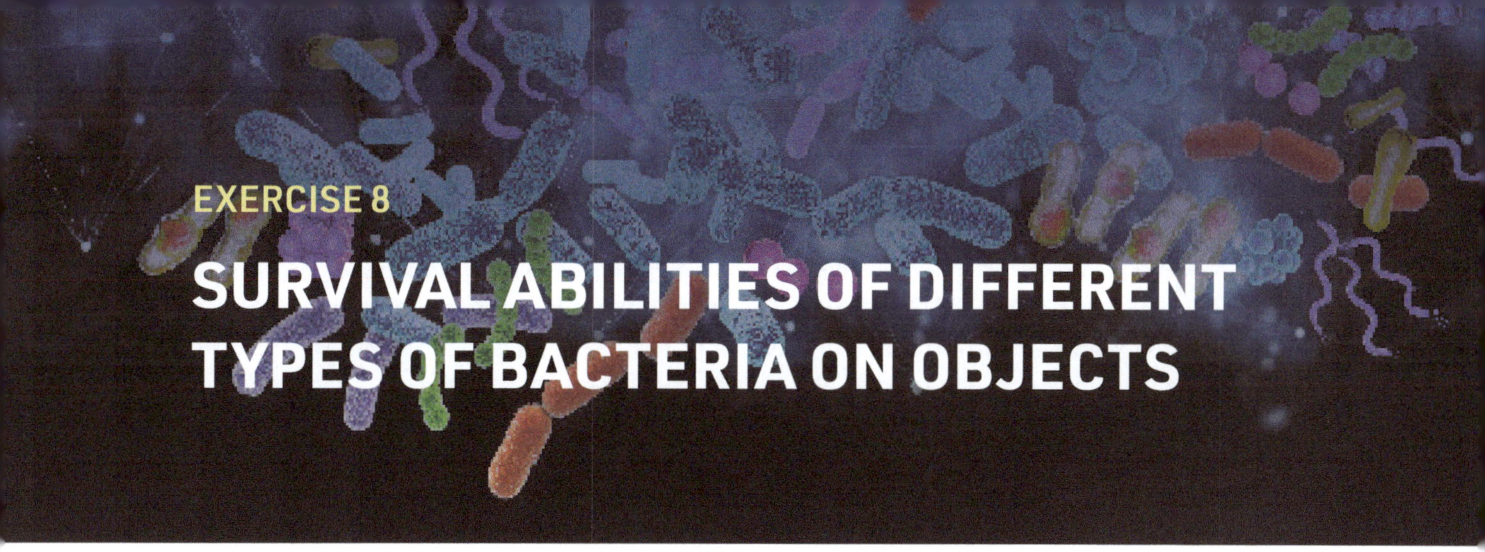

EXERCISE 8

SURVIVAL ABILITIES OF DIFFERENT TYPES OF BACTERIA ON OBJECTS

LEARNING OBJECTIVES

After this exercise, you should be able to

1. Describe the purpose of endospore formation.
2. Explain why *Mycobacterium tuberculosis* is more resistant to antibiotics, digestive enzymes, and immune cell attack than other bacteria.
3. Explain exemplary strategies some bacteria use to survive in an adverse environment.

Introduction

Some microorganisms use unique strategies to survive in an adverse environment. For example, some soil bacteria produce endospores to survive various adverse conditions the soil environment might experience, such as high temperature during the summer season, low temperature during the winter season, UV irradiation, less water, and high salt environment during drought. There are reports that bacterial endospores have survived for thousands of years. Endospores are especially important from a clinical viewpoint because they are resistant to processes that normally kill vegetative cells. Such processes include heating, freezing, desiccation, radiation, and chemical treatment. Whereas most vegetative cells are killed by temperatures above 70°C, endospores can survive in boiling water for several or more hours. Endospore-forming bacteria are also a major problem in the food industry because their endospores are likely to survive under food processing; when they grow, some species produce toxins. During the preparation of canned foods, an important primary process is destroying endospores of *Clostridium botulinum,* which causes botulism. During canned food preparation, food materials are heated inside a can. This heating process drives air out of the can and creates an anaerobic condition. Since *Clostridium botulinum* is a strict anaerobic bacterium, this condition allows the endospore to germinate and grow inside the can. Thus, canned foods should be thoroughly sterilized to kill all endospores.

Another way bacteria protect themselves from harsh environmental conditions and host immune attacks is to produce tough materials and place them on their surface. One example is *Mycobacterium* spp. producing a waxy lipid, mycolic acid. The waxy lipid confers the bacteria more resistance to desiccation, harmful chemicals, antibiotics, digestive enzymes, immune attacks, and the like.

Bacteria exist on almost all surfaces of objects around us, such as surfaces of desks, computers, smartphones, pens, keyboards, papers, and bookshelves. In this exercise, you will determine the survival ability of bacteria when they are on these dried objects. To create a similar environment, we will smear bacteria on the surface of sterile disposable petri dishes. Bacteria will be dried and have their survival ability checked on that environment to examine how much bacteria can survive when they are on the surface of an object.

Materials needed

- Each broth culture (5 or 10 ml) of *Mycobacterium smegmatis*, *Bacillus subtilis*, *Escherichia coli*, and *Staphylococcus aureus* per lab bench
- 8 TSA plates per pair of students
- Sterile cotton swabs
- 37°C incubator

Procedure (pair exercise)

Day 1

1. Divide an empty petri plate into quarters and label them with Ms, Bs, Ec, and Sa as shown in **Figure 8.1** (replace the numbers with microbes' initials).
2. Draw a quarter-sized circle in the center of each quadrant (**Figure 8.1**).

3. Using a sterile cotton swab, make a smear of each bacterium within the circle in each quadrant. (To make a proper smear, the cotton swab should not have too much broth. Squeeze the swab inside the test tube to remove excess broth.)
    a. Quadrant 1: *Mycobacterium smegmatis* (acid-fast bacillus)
    b. Quadrant 2: *Bacillus subtilis* (endospore-forming Gram-positive bacillus)
    c. Quadrant 3: *Escherichia coli* (Gram-negative bacillus)
    d. Quadrant 4: *Staphylococcus aureus* (Gram-positive coccus)
4. Air-dry all the smears.
5. During the drying time, label four TSA plates 1(Ms), 2(Bs), 3(Ec), and 4(Sa). When the smears are dried, swab the smears of *Mycobacterium smegmatis* and spread the organism to a TSA agar plate 1(Ms) using the method below.
    a. Gently brush the swab over the entire surface of the agar.
    b. Rotate the plate 90° degrees and swab the entire surface of the agar plate again.
6. Spread the three other organisms to TSA agar plates with the matching label.
7. Incubate the plate at 37°C until you see bacterial cells on plates. This usually takes one or two days. It might take longer for *Mycobacterium smegmatis* (three to four days).
8. Save the dry smears on the petri dish at room temperature for seven days.

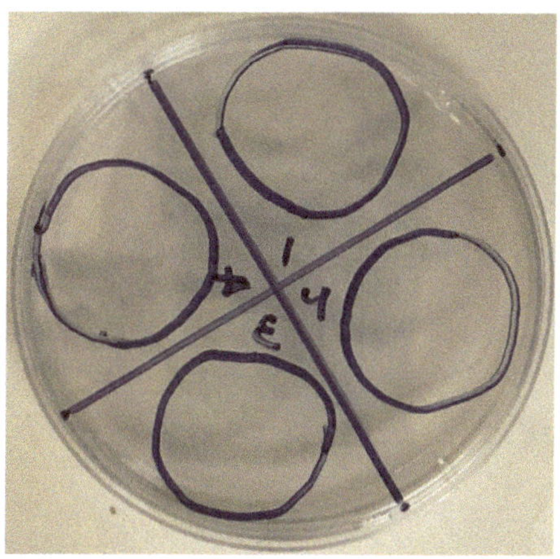

FIGURE 8.1 Preparation of an empty petri dish plate before smearing of each bacterium.

Day 2

1. Record your results in **Table 8.1**.
2. Label four TSA plates Ms-7D, Bs-7D, Ec-7D, and Sa-7D. Spread the smears following the procedure described above (steps 5 and 6).
3. Incubate the plate at 37°C for at least two days. It might take longer for *Mycobacterium smegmatis* (three to four days).

Day 3

1. Record your results in **Table 8.1**.

TABLE 8.1 Survival of microorganisms.

Fill in your results in the blanks using the symbols:

(-): no growth, (+) light growth, (++) moderate growth, (+++) heavy growth

Bacteria	Survival on day 1	Survival on day 7
Mycobacterium smegmatis		
Bacillus subtilis		
Escherichia coli		
Staphylococcus aureus		

Review questions

1. Complete killing of endospores is crucial when canned foods are prepared. Why?

2. What condition does the smearing of bacteria on the surface of the petri dish in this exercise mimic?

3. Which types of bacteria have higher survival ability than others in this exercise? Why?

EXERCISE 9

THE EFFECT OF TEMPERATURE AND OSMOSIS ON MICROBIAL GROWTH

LEARNING OBJECTIVES

After this exercise, you should be able to

1. Define optimum, maximum, and minimum growth temperatures.
2. Define psychrophiles, psychrotrophs, mesophiles, thermophiles, and hyperthermophiles.
3. Classify organisms based on their preferred growth temperatures.
4. Classify organisms based on their osmotic tolerance.
5. Explain what plasmolysis is and when it occurs.

Introduction

Each microbe has a preferred growth temperature range. Microbes grow slower at higher and lower temperatures than their optimum growth temperatures. The microbial growth curve based on temperature is shown in **Figure 9.1**. The growth curve generally shows a bell or cone shape. Their growth gradually decreases when the temperature goes away from the optimum temperature, and eventually, the temperature reaches the minimum or maximum growth temperature.

Based on preferred temperature ranges, microbes are classified as psychrophiles, psychrotrophs, mesophiles, thermophiles, and hyperthermophiles (**Figure 9.2**). The root word *phil* comes from a Greek verb meaning "to love," so psychrophiles, mesophiles, and thermophiles are cold-, moderate-, and

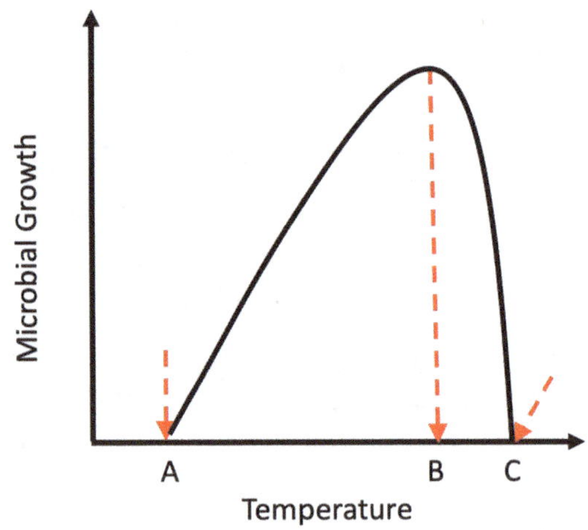

A: Minimum Growth Temperature
B: Optimum Growth Temperature
C: Maximum Growth Temperature

FIGURE 9.1 Microbial growth curve and growth temperatures.

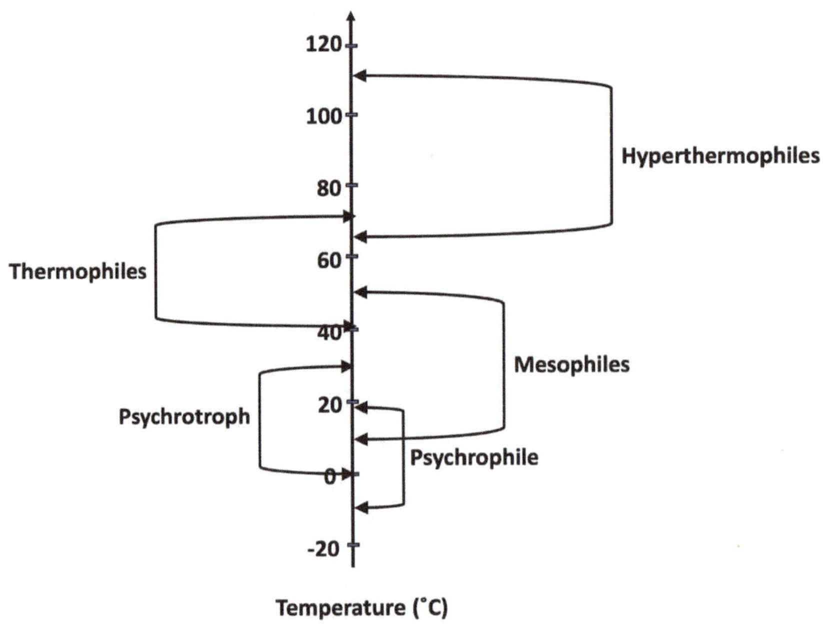

FIGURE 9.2 Microbial classification based on their growth temperature range. The bottom and top arrows of each bracket indicate minimum growth temperature and maximum growth temperature.

high-temperature-loving bacteria, respectively. Psychrophiles can grow below 0°C and at the refrigerator temperature but not the room temperature. On the contrary, psychrotrophs cannot grow below the freezing point and grow between 0°C and 20–30°C. Since psychrotrophs can grow both at refrigerator temperature and room temperature, they can cause food spoilage. Mesophiles have a growth temperature range of 25–40°C. Microorganisms that have adapted to live in the bodies of animals usually have an optimum temperature close to that of their host body temperatures. The optimum temperature for most human pathogenic bacteria is around 37°C. Mesophiles include most of the common food spoilage and disease organisms. Thermophiles are microorganisms capable of growth at high temperatures. Many of these organisms have an optimum growth temperature of 50–60°C, the temperature of water from a hot water tap. Such temperatures can also be reached in sunlit soil and in thermal waters such as hot springs. Many thermophiles cannot grow at temperatures below 45–50°C. Some microbes, such as members of the Archaea called hyperthermophilic archaea, have an optimum growth temperature of 80°C or higher. These organisms are called hyperthermophiles or extreme thermophiles. Most of these organisms live in hot springs associated with volcanic activity.

Cells can't live without water because water is needed for metabolic reactions and to dissolve ions and molecules. Osmotic pressure created by water movements from one place to another is important in microbial growth since it decides the amount of water available in microbes. The high salt or sugar concentrations in the environment cause high osmotic pressure. This hypertonic environment draws water out of microbial cells and causes the separation of cytoplasm from cell walls. This phenomenon is called plasmolysis and inhibits microbial growth or kills microbes. Creating a hypertonic environment is used to preserve certain foods. Salted fish, honey, canned fruits, and sweetened condensed milk are preserved mainly by this mechanism.

Microbes show different tolerance or growth ability in different NaCl concentration solutions (**Table 9.1**). Most microorganisms living around us including microbes in the soil, pond water, river water, and normal microbiota are **facultative halophiles**. Facultative means optional. They do not require high salt concentrations to grow but can grow at salt concentrations up to 2% when they encounter. Microbes that grow in higher than 3% NaCl are **halophiles**. Seawater contains about 3.5% salts, so almost all microorganisms in seawater are halophiles. Microorganisms from the Dead Sea often require nearly 30% salts, and they cannot grow in water with low salt concentrations. These bacteria are called **extreme or obligate halophiles**.

TABLE 9.1 Classification of microbes based on the growth at different salinity.

	Facultative halophiles	Halophiles	Obligate halophiles
Growth at <2% NaCl	+++	- or +	-
Growth at 2% NaCl	+++	++	-
Growth at 5% NaCl	- or +	+++	-
Growth at 10% NaCl	-	- or +	+++

(-) no growth, (+) light growth, (++) moderate growth, (+++) heavy growth

Procedure 1: The effect of temperature on microbial growth (individual exercise)

Materials needed

- Each broth culture (5 or 10 ml) of *Bacillus subtilis, Escherichia coli, Listeria monocytogenes*, and *Geobacillus stearothermophilus* per lab bench
- 1 TSA plate per student
- 1 TSB broth per student
- Refrigerator
- 37°C incubator
- 50°C incubator
- 65°C water bath

Day 1

1. Divide your agar plate into four areas by drawing a cross on the bottom of the plate as shown in **Figure 9.3** and label each area with microbes' initials such as Bs, Ec, Lm, and Gs (replace the numbers with microbes' initials).
2. Using an inoculating loop, inoculate the microorganisms below onto the quadrants by performing a streak as shown in **Figure 9.3**.

a. Quadrant Bc: *Bacillus subtilis* (Gram-positive bacillus, spore-former)
b. Quadrant Ec: *Escherichia coli* (Gram-negative bacillus)
c. Quadrant Lm: *Listeria monocytogenes* (Gram-positive bacillus)
d. Quadrant Gs: *Geobacillus stearothermophilus* (Gram-positive thermophilic bacteria, optimum growth temperature: ~65°C)

3. Place the plate at one of the four different incubation temperatures for 24–48 hours: 4°C (refrigerator), ~22°C (room temperature), 37°C, or 50°C.
4. Inoculate one of the organisms above to a TSB broth aseptically and place into a water bath with temperature of 65°C for 24–48 hours (each student at the same lab bench should inoculate a different microbe).

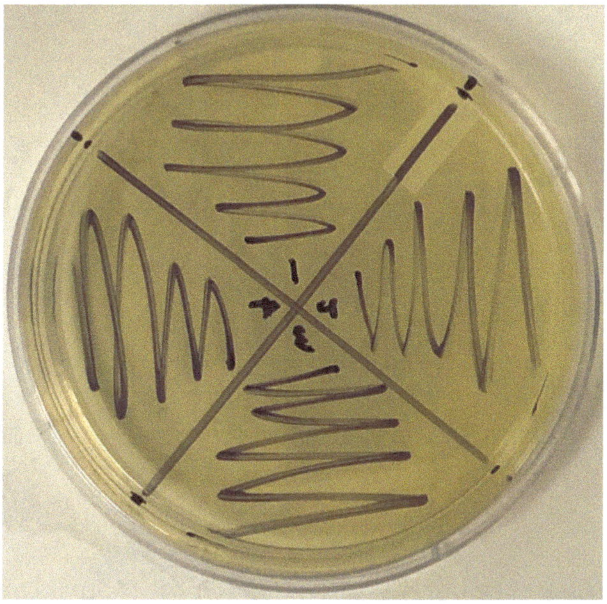

FIGURE 9.3 Streaking of microbes on a plate in this exercise.

Day 2

1. Observe the plates and broth culture and record microorganisms' growth in **Table 9.2**.

TABLE 9.2 Growth of the microbes.

Fill your results in the blanks using the symbols below:

(−) no growth, (+) light growth, (++) moderate growth, (+++) heavy growth

Bacteria	4°C	~22°C*	37°C	50°C	65°C
Bacillus subtilis					
Escherichia coli					
Listeria monocytogenes					
Geobacillus stearothermophilus					

*Room temperature.

Procedure 2: The effect of osmotic pressure on microbial growth (individual exercise)

Materials needed

- Each TSA agar plate streaked with *Staphylococcus aureus*, *E. coli*, and *Listeria monocytogenes*, respectively, per lab bench
- One TSA plate with 2%, 5%, or 10% of NaCl per student.

Day 1

1. Each student will use one TSA plate. Divide the plate into four regions using a marker as shown in **Figure 9.3**.
2. Mark each area with the microbe's initials such as Sa, Ec, Lm, and B (blank). Streak each area of the plate with the corresponding microbe.
3. Incubate the plate at 37°C for 24–48 hrs.

Day 2

1. Fill out the result of the growth of each bacterium in **Table 9.3**.

TABLE 9.3 Growth of microbes at different osmotic pressure (salinity).
(-) no growth, (+) light growth, (++) moderate growth, (+++) heavy growth

	Staphylococcus aureus	*Escherichia coli*	*Listeria monocytogenes*
Growth at 2% NaCl			
Growth at 5% NaCl			
Growth at 10% NaCl			

Review questions

1. What are the minimum, optimum, and maximum growth temperatures of the bacteria you used in this exercise (*Bacillus subtilis*, *Escherichia coli*, *Listeria monocytogenes* and *Geobacillus stearothermophilus*)? You can do internet searches.

Bacteria	Minimum growth temp.	Optimum growth temp.	Maximum growth temp.
Bacillus subtilis			
Escherichia coli			
Listeria monocytogenes			
Geobacillus stearothermophilus			

2. Classify the microbes used in this exercise based on their growth temperature characteristics (psychrophiles, psychrotrophs, mesophiles, thermophiles, or hyperthermophiles).

Bacillus subtilis	
Escherichia coli	
Listeria monocytogenes	
Geobacillus stearothermophilus	

3. Give at least three examples of foods whose shelf lives are increased by using a hypertonic condition.

4. Classify the microbes used in this exercise based on their osmotic tolerance (facultative halophile, halophile, or obligate halophile).

Staphylococcus aureus	
Escherichia coli	
Listeria monocytogenes	

EXERCISE 10

FOOD MICROBIOLOGY

Detection of Microbial Contamination in Foods

LEARNING OBJECTIVES

After this exercise, you should be able to

1. Explain the difference between direct methods and indirect methods to determine the degree of microbial contamination in foods.
2. Perform how to measure microbial contamination in foods.
3. Explain the role of methylene blue in determining microbial contamination in foods.
4. Explain why hamburger patties should be thoroughly cooked, not like steak.

Introduction

Food materials have lots of nutrients that can support microbial growth. If foods are not stored properly, they will be easily spoiled by microorganisms. So the food industry needs a way to accurately examine and monitor microbial contamination in foods. Microbial growth in foods usually indicates the spoilage of foods, other than fermented foods such as kimchi, sauerkraut, and yogurt, so determining microbial growth is necessary to determine the degree of food spoilage. Microbial growth in foods can be determined directly or indirectly. Microbial cell numbers can be counted by counting colony-forming units on plates (plate count) or counting cells under a microscope (direct microscopic count). Microorganisms consume or produce some chemicals by enzymatic reactions. Measuring the metabolic activities, microorganism numbers can be estimated indirectly by quantitating produced or missing chemicals. In this exercise, you will practice both direct and indirect counting methods.

The plate count method counts colony numbers on agar plates. First, samples should be properly diluted and spread on agar plates to obtain countable colonies. Then, the agar plates should be incubated at proper conditions for an adequate amount of time (e.g., plates are incubated at 37°C overnight). During the incubation, cells multiply and form a visible group of cells called a colony. Plate counts assume that each live bacterium grows and divides to produce a single colony. However, this is not always true because bacteria frequently grow linked in chains or clumps such as the case of *Streptococci* or *Staphylococci*. Therefore, a colony often results not from a single bacterium but rather from a bacterial chain or clump. Thus, a colony is a population of cells arising from a single cell or a group of attached cells. Plate counts are often reported as colony-forming units (CFUs) to reflect this reality. When there are 100 colonies on a plate, rather than describing it as 100 cells in the original culture, it should be stated as 100 CFUs in the original culture. The plate count method is the most frequently used method for measuring bacterial populations. Plate counts only measure the number of viable cells since dead cells do not multiply. This method needs time to grow microbial cells until they form visible colonies. Usually, it takes more than 24 hours. This can be a problem in some applications such as in the quality control of milk.

To have agar plates with properly countable colonies, serial dilutions are used. In serial dilutions, the original sample is diluted in a series of dilution tubes. For example, each succeeding dilution tube will have only one-tenth the number of microbial cells as the preceding tube (**Figure 10.1**). Then, the diluted samples are used to inoculate agar plates, on which colonies grow and can be counted. This count is then used to estimate the number of bacteria in the original sample. After the serial dilution, the plates with colony numbers between 25 to 250 are used for counting. If there are too many colonies on a plate, they cannot be counted accurately because some colonies overlap each other. If there are too few colonies, it can give big errors. For example, on the 1 to 100,000 dilution plate in **Figure 10.1**, a colony represents 100,000 CFUs.

Meats and poultry can be contaminated by microorganisms during the butchering process. In most cases, the internal parts of meats are usually sterile, and only the surface of meats is contaminated (e.g., steaks or chicken breasts). These microorganisms on the meat surfaces are killed during cooking. However, if meat is ground, the surface microorganisms can be internalized inside meat patties. Thus, meat patties should be thoroughly cooked, otherwise, they could cause food poisoning. This is why fast-food restaurants using hamburger patties are prone to causing food poisoning outbreaks.

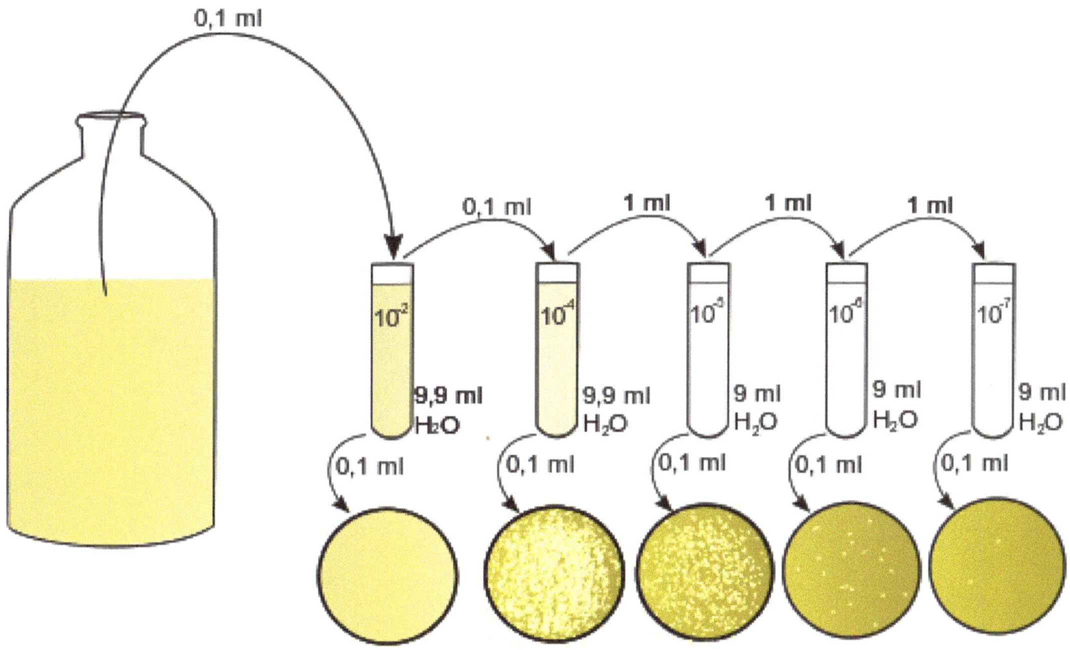

FIGURE 10.1 An example of serial dilution of a microbial sample. The sample is diluted serially to 10^2, 10^4, 10^5, 10^6, and 10^7 times.

There are many indirect methods using microbial metabolic activities to check microbial contamination in foods. In the milk industry, measuring microbial contamination should be quick and easy since microorganisms can grow quickly in a nutrient-rich environment. Thus, the plate-counting method is inappropriate. One of the popular indirect methods to measure how many microorganisms present in milk is measuring biological oxygen demand (BOD). Most milk-spoiling microorganisms in milk need oxygen to grow and this oxygen demand is proportional to the number of microorganisms. BOD can be determined using a chemical that changes its properties after binding to oxygen. One of the chemicals is methylene blue, which changes from blue to colorless when it loses oxygen. When methylene blue is added to milk, its color is blue since milk contains some oxygen. When microorganisms grow during incubation, the microorganisms consume oxygen, and methylene blue turns colorless. The speed of the color change can indicate how many microorganisms are present in the milk initially.

Procedure 1: Detection of microbial contamination in milk (individual exercise)

Materials needed

- One carton of pasteurized milk stored at 4°C per lab bench
- One carton of pasteurized milk stored at 37°C overnight per lab bench
- Two sterilized tubes with caps per student
- Two 1 ml pipettes per student
- Methylene blue (4 parts per million, ppm)
- Procedure in Header weight

1. Label the two empty tubes as 4 or 37.
2. Add 9.0 ml milk from a refrigerator to the tube labeled as 4.
3. Add 9.0 ml milk incubated at 37°C to the tube labeled as 37.
4. Add 1.0 ml methylene blue to each tube using 1 ml pipettes and gently mix with the pipettes. Do NOT introduce air into milk as much as possible. Do not vortex.
5. Incubate them at 37°C without shaking in a water bath. Check the color of the tubes every 30 minutes for 120 minutes total.
6. Record the result in **Table 10.1**.

TABLE 10.1 The time required for color change.* Mark the time when the color changes.

	Incubation time (min)			
	30	**60**	**90**	**120**
Refrigerated milk				
Milk stored at 37°C overnight				

*The faster the blue color disappears, the higher the bacterial count. Unspoiled milk should remain blue for at least 6 hours.

Procedure 2: Detection of microbial contamination in hamburgers (individual exercise)

Materials needed

- Hamburger patty
- 1 TSA plate per student

- 1 EMB plate per student
- Cell spreaders (aka hockey sticks): This is the first time that students use cell spreaders
- 95% alcohol in a beaker

Day 1

1. Place 10 grams of hamburger in a blender.
2. Add 90 ml of sterile water and blend (~10 times dilution). This can be used by all students in the lab. A student or the instructor can do this.
3. Wait until most solid materials are precipitated.
4. Take 0.1 ml (100 µl) of the supernatant and add it in the middle of a TSA and EMB agar plate each. This is equivalent to 1,000X dilution (100 ml to 0.1 ml).
5. Spread the sample over the entire surface of the plate using the cell spreader.
    a. Soak the spreader head in the ethanol solution.
    b. Take the spreader out of the ethanol solution and remove excess ethanol.
    c. Flame the spreader head.
    d. When the flame of the spreader head goes out, cool down the spreader by touching on fresh agar on an agar plate.
    e. Spread the sample evenly all over the plate.
    f. After spreading, put the spreader head in the ethanol solution.
6. Incubate the plates at 37°C degree for 24–48 hours.

Day 2

1. Count the colonies on plates that have between 25 and 250 colonies.
2. Calculate the total number of microorganisms per gram of meat, considering the sample dilution rate.
3. Record your result and the section average value in **Table 10.2**. The TA will collect your result to calculate the average value of the students' results in the section.
4. Perform a Gram stain with the dominant colony from each plate (each student will perform one Gram staining).

TABLE 10.2 Colony characteristics and abundance of microorganisms in hamburger patty.

Media	CFUs/ plate[1]	CFUs per gram of hamburger	# of colony types on the plate	Gram type (+ or –)[2]	Shape of the bacterium;[3] draw the shape below
TSA					
EMB					

[1]CFU: Colony Forming Unit
[2]Gram type (+ or –) of the bacterium you have chosen
[3]Shape of the bacterium you have chosen under a microscope (coccus, bacillus, streptococcus, staphylococcus, etc.)

- Write down the section average colony number per gram of hamburger patty.

Review questions

1. Why are the plates with colony number between 25 and 250 used to calculate initial bacterial CFUs in samples?

2. What type of microorganisms are dominant in contaminated hamburgers?

3. After mixing milk and methylene blue, this solution should not be vortexed. Why?

4. Why does the top portion of all the tubes tend to remain blue longer?

5. Is refrigeration an effective short-term storage method for milk?

6. You do not want to choose a plate with too few colonies (e.g., 10 colonies) to count bacterial cell number in a sample after serial dilution. Why? Please explain in detail.

Credit

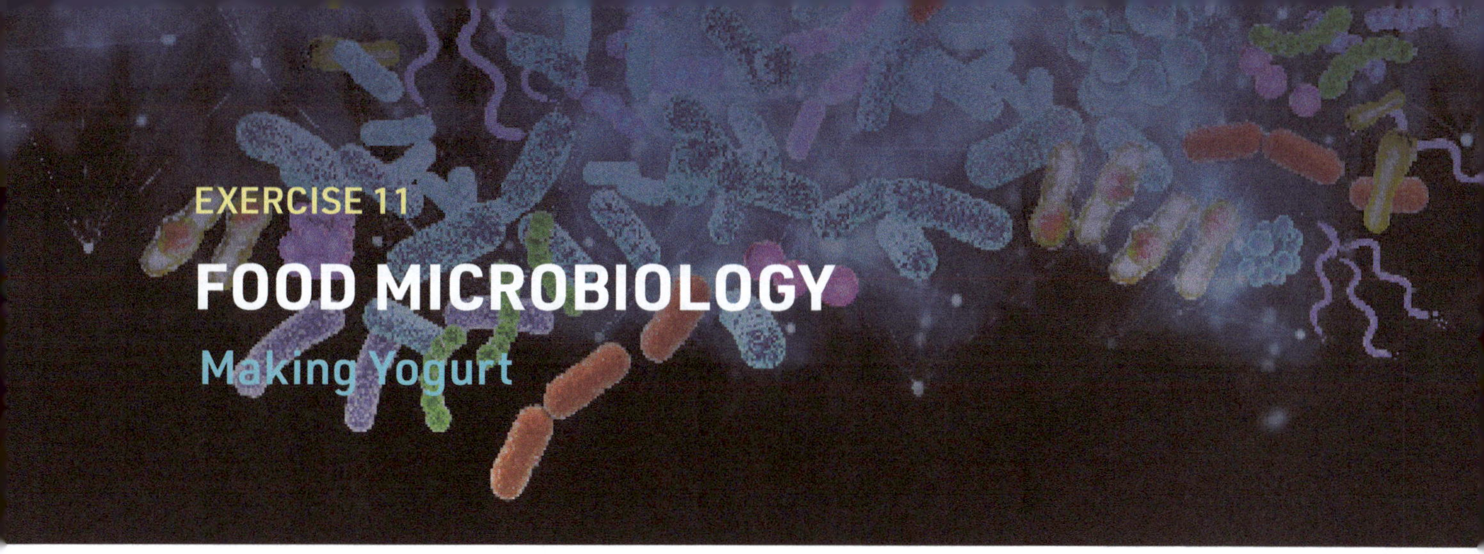

EXERCISE 11

FOOD MICROBIOLOGY

Making Yogurt

LEARNING OBJECTIVES

After this exercise, you should be able to

1. Explain what benefit fermented foods provide, compared to raw foods.
2. Tell what microorganisms are used to make yogurt.
3. Know which ones are crucial steps when you make yogurt.
4. Understand why pure cultures are used to make yogurt.

Introduction

People in different areas of the world have developed diverse fermented foods. When certain foods are fermented, they usually have longer shelf lives and favorable flavors compared to raw foods. Examples of popular fermented foods are kimchi, sauerkraut, kombucha, cheese, and yogurts. Most foods are fermented by acid-producing bacteria that can serve as probiotics, making our digestive system healthy. When these acid-producing bacteria grow, they lower the pH of food and change flavor. Also, these bacteria change texture by digesting some components in the foods, providing better taste and making them easily digestible.

In this exercise, you will practice making one of the most popular fermented foods: yogurts. To make fermented foods, usually normal microflora that are naturally associated with the raw materials have been used. To prevent unwanted microbial contamination by food spoiling or pathogenic microorganisms during commercial yogurt fermentation, three key steps are applied. First, pasteurized milk is used.

Pasteurization is done by heating foods at 63°C for 30 minutes. Since pasteurization can kill many microorganisms without significantly changing milk texture, pasteurized milk has significantly fewer food-spoiling and pathogenic microorganisms. Second, the fermentation is done at between 40 to 50°C. The optimum growth temperature of food-spoiling and pathogenic microorganisms is body temperature (37°C) or below. However, some lactic acid bacteria are heat resistant and grow well around 45°C. Third, pure cultures are used to inoculate for fermentation. Pure cultures can be obtained through applying the pure culture technique you learned previously. The microorganisms used to make commercial yogurts include *Streptococcus thermophilus, Lactobacillus bulgaricus, Lactobacillus acidophilus,* and *Bifidobacterium* spp. In this exercise, commercial yogurts will replace the pure culture inoculum. Also, brief boiling of milk will be done before inoculation even though you will use pasteurized milk. For commercial yogurt production, pasteurization is done right before the inoculation of microorganisms. Otherwise, microorganisms in the pasteurized milk will grow when they are exposed to room temperature or higher. This is why you should keep milk in a refrigerator at home.

Materials needed

- One small yogurt per lab bench. Provide diverse yogurts to the class. They should be refrigerated before being used in the lab
- 100 ml pasteurized milk per pair of students

Procedure (pair exercise)

Day 1

1. Heat 100 ml milk to boiling. Remove milk from heat as soon as it boils.
2. Cool the milk down to 50°C. The milk will be cool enough when you can hold the beaker in your hand without burning.
3. Add one tablespoonful of yogurt.
4. Determine the initial pH of the mixture using a pH paper stick.
5. Incubate at 45°C without shaking for 24–48 hours.

Day 2

1. Determine the final pH of the mixture using a pH paper stick.
2. Perform a Gram stain and record your observation in **Table 11.1.**

TABLE 11.1 pH change of the yogurts and morphology of the bacteria in yogurts.

Yogurt pH		Draw the morphology of bacteria under a microscope. Are the bacteria Gram positive or negative?
Initial pH	Final pH	

Review questions

1. What was the name of the yogurt you used? What is the active culture (microorganism) in the yogurt?

2. How did the pH of yogurt change?

3. What methods were used to prevent microbial contamination when you made yogurt?

EXERCISE 12

PHYSICAL CONTROL OF MICROBIAL GROWTH

LEARNING OBJECTIVES

After this exercise, you should be able to

1. Define the growth of microorganisms.
2. Define what is autoclaving and how to achieve it.
3. Explain how to sterilize liquid or gas through filtration.
4. Explain when UV can be used to control microbial growth.

Introduction

The definition of microbial growth differs from that of animal or plant growth. While the growth of animals refers to an increase in size, microbes do not significantly increase in size. Microbes typically grow to less than twice their original size before dividing. Thus, the growth of microbes is primarily characterized by rapid division and the production of daughter cells, resulting in an increase in cell number or population size. It is crucial to control the growth of microorganisms to prevent food spoilage and tissue infection. To control microbial growth, it is necessary to target critical cellular structures or functions. Antimicrobial agents can damage various components of microbial cells, such as the plasma membrane, proteins, or nucleic acids, all of which play essential roles in cellular activities. In this exercise, you will learn how to kill microorganisms or restrict microbial growth using physical methods.

One of the most popular methods of controlling microbial growth is by using heat. Heat can denature and coagulate proteins, which ultimately destroys their structures and functions. Protein denaturation and coagulation can be observed when cooking eggs, as the egg white becomes opaque and does not return to its original transparent state even after cooling. Heat can be classified into two categories: moist heat and dry heat. Moist heat is more efficient in killing microbes, as heat can be delivered directly into microbes through water vapor. Boiling an object for 30 minutes can kill most vegetative microbes, which is why boiling kitchen towels is a common method of disinfection. However, boiling water cannot kill bacterial endospores, which are the toughest microbial form to kill. To achieve sterilization (killing all microorganisms, including endospores), the temperature must be increased beyond boiling point (100°C). The most used temperature for sterilization is 121°C, which can kill endospores when exposed for at least 15 minutes. If the object being sterilized is large, the incubation time should be increased to ensure that heat reaches the center of the object. The autoclaving condition is 121°C for 15 minutes, steam must be pressured at 15 psi. Pressure cookers can also be used to sterilize items at home, as they generate pressured steam that can increase the inside temperature up to 121°C. In hospitals and research labs, machines called autoclaves are used to sterilize various items such as medical instruments, culture media, and microbe-contaminated materials (**Figure 12.1**). Essentially, autoclaves are professional-grade pressure cookers with enhanced safety features to ensure fail-proof operation.

FIGURE 12.1 Autoclaves.

Proper filtration can effectively remove microbes from liquid or gaseous samples. Filtration works by forcing the liquid or gas through a filter with very small pores to trap and eliminate microbes. Typically, the filter pores have diameters of less than 0.5 microns, which are smaller than most microbes except for viruses. Filtration is an efficient and cost-effective method to sterilize air because the filters do not need to be replaced frequently. High-efficiency particulate air (HEPA) filters with pore sizes of less than 0.3 microns are used to remove microbes from the air. These days, most vacuum cleaners have a HEPA filter to trap allergens and microbes. For liquid sterilization, membrane filters are used (**Figure 12.2**). Since membrane filtration can be more expensive than autoclaving, it is only used for liquid materials sensitive to heat, such as antibiotics or vitamin solutions. The pore size of membrane filters for liquids is typically 0.22 microns.

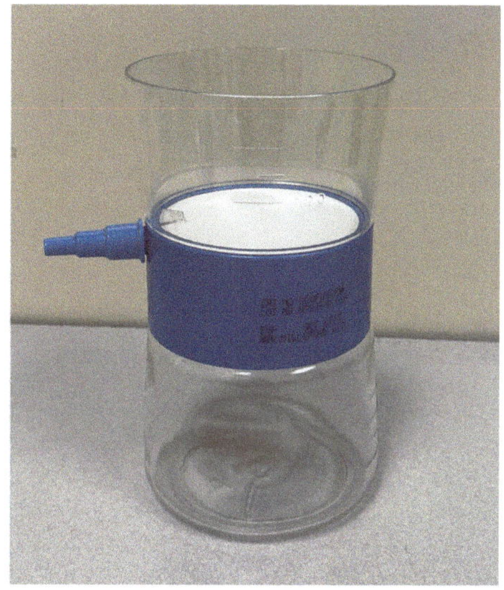

FIGURE 12.2 A liquid filtration kit.

UV radiation is another method used for controlling microbes. Germicidal UV lamps are commonly found in hospital rooms, nurseries, operating rooms, and cafeterias. **Figure 12.3** shows a germicidal UV lamp inside a cell culture cabinet. Since UV radiation does not have strong penetration power, the target object that requires disinfection should be directly exposed to UV light.

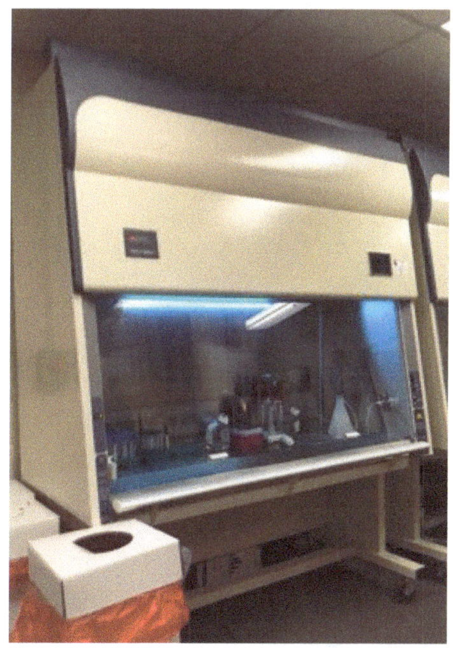

FIGURE 12.3 A germicidal UV lamp inside a cell culture cabinet.

Procedure 1: Autoclave (group exercise)

Materials needed

- Four TSB broth tubes per lab bench
- One tube of *Bacillus subtilis* broth culture per lab bench

- Pans to boil water
- 70°C water bath

Day 1

1. Get four TSA broth tubes from the instructor.
2. Label the tubes: A, B, C, and 70.
3. Inoculate a full loopful of *Bacillus subtilis* culture into the four TSA broth tubes.
4. Provide the treatment to each tube.

   Tube A: **a**utoclave
   Tube B: incubate it inside **b**oiling water for 10 minutes
   Tube C: **c**ontrol, no treatment
   Tube 70: incubate in a **70**°C water bath for 10 minutes
5. The instructor will show the autoclave machine and explain how to use it.
6. Incubate the tubes at 37°C **for 24** hours and store at 4°C.

Day 2

1. Observe and record the growth of the microorganism in **Table 12.1**.

TABLE 12.1 Survival of microorganisms.

Fill in your results in the blanks using the symbols:

(-) no growth, (+) light growth, (++) moderate growth, (+++) heavy growth

Bacterium	A: autoclaved	B: boiled for 10 min	70: incubated at 70°C for 10 min	C: no treatment
Bacillus subtilis				

Procedure 2: Filtration (individual exercise)

Materials needed

- One broth culture of *Escherichia coli* (per lab bench)
- 100 ml sterile water (per lab bench)
- 2 TSA plates (per student)
- Sterile cotton swabs

- Sterile disposable 10 μl inoculating loops (instead, you can use the wire inoculating loop)
- 37°C incubator

Day 1

1. Label the two TSA plates "unfiltered" and "filtered" at the bottom of the plates.
2. Add a loopful (~10 μl) of *E. coli* broth into 100 ml sterile water.
3. Take 100 μl of the inoculated water sample and spread it using a sterile cotton swab on a TSA plate labeled "unfiltered."
4. Using the vacuum tap on the lab bench and filtration apparatus, filter the 100 ml water sample inoculated with *E. coli*. The instructor will demonstrate how to set up the filtration apparatus.
5. Take 100 μl of the filtered water and spread it using a sterile cotton swab on a TSA plate labeled "filtered."
6. Incubate the plates in the 37°C incubator for 24–48 hours and store them in the 4°C refrigerator.

Day 2

1. Observe your plates and fill in the blanks in **Table 12.2**.

TABLE 12.2 Presence of microorganisms.

Fill your results in the blanks using the symbols below:

(-) no growth, (+) light growth, (++) moderate growth, (+++) heavy growth

Bacteria	Unfiltered sample	Filtered sample
E. coli		

Procedure 3: UV radiation (individual exercise)

Materials needed

- Broth cultures of *Mycobacterium smegmatis, Bacillus subtilis, Escherichia coli*, and *Staphylococcus aureus* (per lab bench)
- 2 TSA plates per student
- Sterile cotton swabs

- 37°C incubator
- UV light (if a UV crosslinker chamber is used, use the energy level of 5000 X 100 μJ/cm^2)

Day 1

1. Each student will pick one microorganism.
2. Label your plates

   Plate A: your initials, microorganism initials, UV

   Plate B: your initials, microorganism initials, control
3. Add 0.1 ml of the culture you chose on the center of each plate and spread the bacterium using a sterile cotton swab on the whole surface evenly.
4. Your instructor will demonstrate how to use the UV lamp or the UV cross-linker chamber. Do NOT look at the lamp directly and be careful to minimize your exposure. Expose plate A to UV for a specific time (1 hour under the UV lamp or 5 minutes in the UV crosslinker). When the plates are exposed to UV, the lid should be off.

Label	Lid	Light source
UV	Off	UV light
Control	No treatment	

5. Incubate all plates at 37°C for 24–48 hours and store them at 4°C.

Day 2

1. Share your result with your peers and record the results in **Table 12.3**.

TABLE 12.3 Survival of microorganisms.

Fill in the results in the blanks using the symbols below:

(-) no growth, (+) light growth, (++) moderate growth, (+++) heavy growth

Bacteria	UV	Control
Mycobacterium smegmatis		
Bacillus subtilis		
Escherichia coli		
Staphylococcus aureus		

Review questions

1. Is boiling an object at 100°C for 30 minutes a sterilization method? Why?

2. What is the condition of autoclaving and how is it achieved?

3. For the liquid filtration, what was the pore size of the membrane filter used in the classroom? What is the size of *E. coli*?

4. Which microorganisms that are used in this exercise are the most and least sensitive to UV irradiation? Can you explain why the microorganisms show different sensitivity to UV irradiation?

EXERCISE 13

CHEMICAL CONTROL OF MICROBIAL GROWTH

Antiseptics and Disinfectants

LEARNING OBJECTIVES

After this exercise, you should be able to

1. Distinguish between sterilization, disinfection, antisepsis, and degermation.
2. Explain why phenol is the gold standard of disinfectant.
3. Explain the characteristics of o-phenylphenol, betadine, bleach, and alcohols as antiseptics or disinfectants.
4. Perform the disk diffusion assay.

Introduction

Sterilization is the process of destroying all viable microbes, including viruses and endospores. This process is essential in medical and laboratory settings to prevent the spread of infectious diseases. Disinfection, antisepsis, and degermation, on the other hand, are aimed at reducing the number of microbes, mainly vegetative microbes, without necessarily killing all microbes. Disinfection is the process of destroying mainly pathogens from inanimate objects, but it does not kill endospores. A common example of disinfection is applying a 5% bleach solution to an examining table. Disinfectants are chemical reagents used to achieve this process. Lysol, bleach, and alcohol are examples of disinfectants used for this purpose. Antisepsis is the process of applying chemicals that kill microbes directly to exposed body surfaces, such as wounded tissues. Antiseptic agents are used for this process, and examples include betadine and alcohol. Degermation is the process of reducing the number of microbes from living tissues in a limited area through cleansing techniques, such as surgical hand scrub. This process is medically related and more serious than sanitization. Degerming agents are used to achieve this process.

Phenol was a widely used antiseptic and disinfectant in the past because it kills microorganisms by disrupting the plasma membrane of microorganisms. However, due to its skin irritation, unpleasant odor, and availability of better options, phenol is not frequently used nowadays. Despite this, phenol is still considered the standard for disinfectants, and other disinfectants are often compared to it to determine their efficacy. Cresol, a phenolic compound, is commonly used as an antiseptic these days. O-phenylphenol, a derivative of cresol, is an active component in many formulations of disinfectants.

Halogens like iodine and chlorine are commonly employed as antiseptics or disinfectants due to their effectiveness against various microorganisms. Iodine, for instance, works by disrupting cell membranes and protein synthesis, and it exhibits activity against bacteria, fungi, and some endospores and viruses. Iodine is used as a tincture, or an iodophor. Tinctures are medicine in aqueous alcohol, and iodophor is a compound that slowly releases iodine from organic molecules. Betadine is a widely used iodophor with the antimicrobial activity of iodine but is less irritating and does not stain, compared to iodine tincture.

Chlorine is a highly effective disinfectant widely used in various applications. It is used to disinfect tap water, swimming pool water, and sewage due to its strong oxidizing properties. The most used disinfectant, bleach, is a chlorine derivative called hypochlorous acid. Chloramines are another group of chlorine compounds that form when chlorine reacts with ammonia or amines in water. Most municipal water treatment systems add ammonia to chlorine to form chloramines. Chloramines are toxic to aquarium fish, so if you want to use tap water for your aquarium fish, it is recommended that you do not use it immediately. Instead, you should store the tap water at room temperature for a day or two or use chemicals sold by pet shops to remove or neutralize chloramines.

Alcohols, particularly ethanol and isopropanol, are commonly used as disinfectants due to their germicidal properties. They are widely used in healthcare settings, and many hand sanitizers contain either ethanol or isopropanol. The antimicrobial activity of alcohols is due to their ability to denature and coagulate proteins. Alcohols also dissolve membrane lipids, which further contributes to their effectiveness as disinfectants. Pure ethanol is less effective than aqueous solutions that contain ethanol mixed with water because protein denaturation requires water. Therefore, a solution of 70% alcohol is commonly used for disinfection. Alcohols are effective against a wide range of microorganisms, including bacteria, fungi, and enveloped viruses. However, it is important to note that alcohols may not be effective against all types of microorganisms, such as bacterial spores.

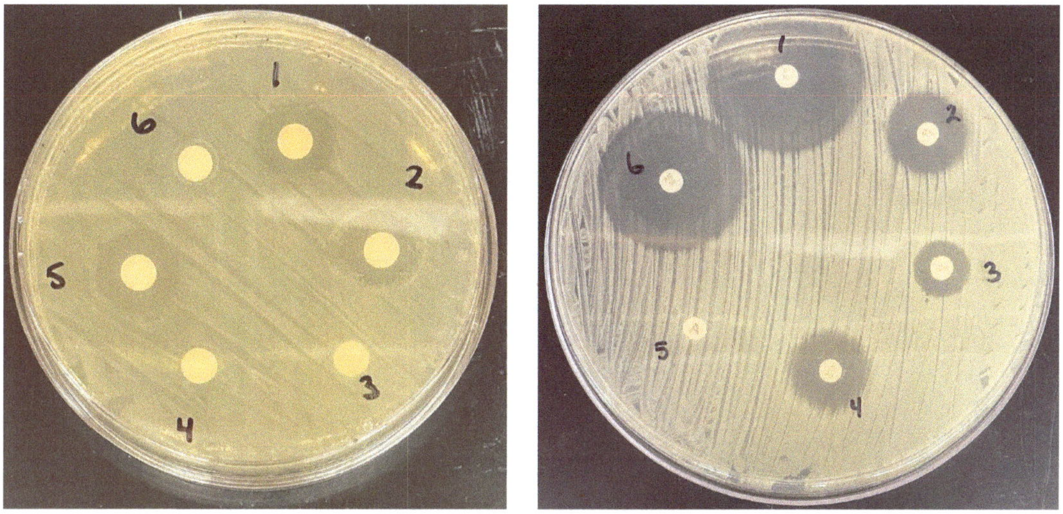

FIGURE 13.1 Examples of the disk-diffusion assay with antiseptics and disinfectant (left) and antibiotics (right).

The disk-diffusion method is a common approach for determining the efficacy of chemical agents against microorganisms. In this method, a disk of filter paper containing a chemical agent is placed on an agar plate previously inoculated with the microbe of interest. Following incubation, a clear zone of inhibition around the disk indicates the antimicrobial efficacy of the chemical agent (**Figure 13.1**). This technique is also utilized for testing the susceptibility of a microbe against diverse antibiotics. The diameter of the clear zone corresponds to the degree of antimicrobial activity of the chemical agent.

Procedure: Antiseptics and disinfectants (individual exercise)

Materials needed

- Liquid cultures of *Escherichia coli*, *Pseudomonas aeruginosa* and *Staphylococcus aureus* per lab bench
- 5 different disinfectants (5 ml each) in beakers per lab bench
- 1 beaker with 95% ethanol for sterilization per lab bench
- 1 TSA plate per student
- Forceps
- Sterile cotton swabs

- Filter paper disks
- Rulers

Day 1

The instructor will provide antiseptics and disinfectants below.

Number	Chemical name
1	Betadine®
2	Isopropyl alcohol
3	Listerine®
4	Lysol®
5	10% bleach
6	Sterile water (negative control)

1. Each student will use one bacterium and all the six solutions above. Choose one among the bacteria provided.
2. Using a sterile cotton swab and spread the bacterium evenly on a TSA plate to make a lawn of bacteria. This can be achieved by swabbing the entire surface of the plate three times, rotating the plate 90° first and then 45° the second time. Discard swabs in a biohazard bag.
3. Label your plate with your initials and the initials of the bacterium you used (e.g., EC, PA, or SA).
4. Mark 1 to 6 on the bottom of the plate like **Figure 13.2**.
5. Flame-sterilize the forceps.
   a. Place the forceps in the beaker of 95% ethanol briefly.
   b. Flame the forceps using a Bunsen burner flame. You are not heating the forceps, just igniting the alcohol.
   c. Hold the forceps in the air until all the alcohol has burned off.
6. Cool down the forceps in the air and pick up one filter paper disk in the glass petri dish.
7. Touch the disk on the surface of the disinfectant #1, which is betadine solution, then the disinfectant will soak into the disk. Do not submerge the disk in the solution.
8. Place the disk filter on the place of the TSA agar plate where you labeled 1.
9. Tap it down gently with the forceps to ensure that it adheres to the surface of the agar.

10. Repeat steps 5–9 for all solutions.
11. Incubate the plates for 24–48 hours, and store at 4°C.

Day 2

1. Record your results and those of your peer students in **Table 13.1**.

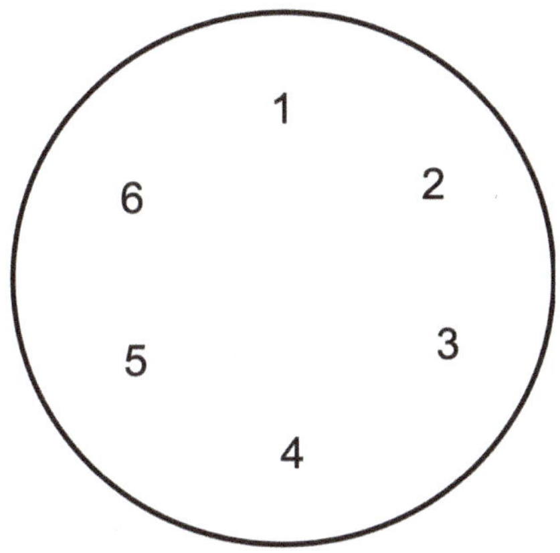

FIGURE 13.2 Mark the numbers on the bottom of your TSA plate.

TABLE 13.1 The diameters of clear zones (mm) caused by antiseptics or disinfectants.

Number	Chemical name	*Escherichia coli*	*Pseudomonas aeruginosa*	*Staphylococcus aureus*
1	Betadine®			
2	Isopropyl alcohol			
3	Listerine®			
4	Lysol®			
5	10% bleach			
6	Sterile water			

Review questions

1. Explain the difference between antiseptics and disinfectants.

2. Write down the active ingredients and concentrations of the chemicals you used. Are they antiseptics or disinfectants?

Number	Chemical name	Antiseptic or disinfectant	Active ingredient and its concentration
1	Betadine®		
2	Isopropyl alcohol		
3	Listerine®		
4	Lysol®		
5	10% bleach		

3. Interpret the results in **Table 13.1**. To which chemicals is each organism most resistant or sensitive?

	Most resistant	Most sensitive
Escherichia coli		
Pseudomonas aeruginosa		
Staphylococcus aureus		

4. Overall, which microorganism among the ones you used is most resistant to chemicals?

EXERCISE 14

CHEMICAL CONTROL OF MICROBIAL GROWTH

Antibiotics

LEARNING OBJECTIVES

After this exercise, you should be able to

1. Explain the selective toxicity and the spectrum of antibiotics.
2. Explain the advantages and disadvantages of broad-spectrum antibiotics.
3. List the common microbial targets of antibiotics.
4. Summarize the characteristics of each antibiotic you learn in this exercise.
5. Describe how antibiotic resistance occurs and spreads.

Introduction

Antimicrobial drugs, including antibiotics, are chemotherapeutic agents that interfere with microbial growth within a host. They are selectively toxic to microorganisms and should not harm human cells. These drugs act by interfering with the essential biological processes required for microbial survival. For example, drugs that inhibit cell wall biosynthesis, protein synthesis, membrane function, nucleic acid synthesis, and metabolic pathways can be used to kill microorganisms or inhibit their growth. It is important to use these drugs appropriately and monitor their effects to prevent adverse reactions. They can interact with other drugs to increase their toxicity or alter the microbiome to cause infection by organisms that are normally saprophytic.

Antibiotics are a class of drugs that are used to treat bacterial infections. They were first discovered by Scottish physician and microbiologist Sir Alexander Fleming in 1928, who discovered the antibiotic substance penicillin. Prior to the discovery of antibiotics, infectious diseases were a major cause of death, with one in three children in the United States expected to die of an infectious disease before the age of five. Antibiotics have played a crucial role in reducing the incidence of infectious diseases and improving overall health outcomes. However, bacterial antibiotic resistance is becoming a major concern, as some infections by MRSA (methicillin-resistant *Staphylococcus aureus*) or CRE (carbapenem-resistant *Enterobacteriaceae*) are becoming increasingly difficult to treat. This can lead to life-threatening situations, as even minor infections can become deadly. Therefore, it is important to use antibiotics judiciously and only when necessary, in order to slow the development of antibiotic resistance and preserve their effectiveness for future generations.

Antibiotics can be classified into two main types based on their mechanism of action. Bactericidal antibiotics kill bacteria directly, while bacteriostatic antibiotics only inhibit bacterial growth. When bactericidal antibiotics are used to treat bacterial infections, they kill the bacteria completely, and removing the antibiotics from the bacterial culture after overnight treatment would not allow the bacteria to grow again. However, when bacteriostatic antibiotics are used to treat bacterial infections, the bacteria only stop growing, and removing the antibiotics from the bacterial culture after overnight treatment would allow the bacteria to grow again. Many antibiotics are bacteriostatic, including antibiotics that inhibit protein synthesis by binding to ribosomes. Although bacteriostatic antibiotics do not kill bacteria directly, they are still effective for bacterial infections because the host's own immune system can clear the bacteria from the body when the bacteria do not grow.

Antibiotics are classified based on their activities toward the spectrum of microbes: broad-spectrum antibiotics and narrow-spectrum antibiotics. Broad-spectrum antibiotics can affect a broad range of microbes, including both Gram-positive and Gram-negative bacteria. They are useful in treating infections when the pathogen causing the infection is not identified quickly. However, these antibiotics can also destroy normal microbiota that is critical to maintaining a healthy body, leading to adverse effects such as superinfection. For example, superinfection occurs when a broad-spectrum antibiotic kills healthy intestinal microflora, causing *Clostridium difficile*, a naturally existing microbe in the intestine and resistant to many antibiotics, to flourish and cause gut troubles such as colitis. Tetracycline kills almost all bacteria including intracellular pathogens such as *Chlamydia* and *Rickettsia*, so it is a broad-spectrum antibiotic and notorious for causing superinfection. Narrow-spectrum antibiotics, on the other hand, have a narrower range of microbial types they affect. They target specific groups of bacteria and are less likely to destroy normal microbiota than broad-spectrum antibiotics. Some narrow-spectrum antibiotics kill only Gram-negative bacteria, and some kill only Gram-positive bacteria. For example, penicillin G kills Gram-positive bacteria but not Gram-negative bacteria because it cannot penetrate the outer membrane of Gram-negative bacteria. It is generally recommended to use narrow-spectrum antibiotics when possible, to minimize the destruction of normal microbiota and the development of antibiotic resistance.

To effectively kill or inhibit the growth of bacteria, antibiotics interfere with key structures or processes necessary for the life of bacteria. These processes include cell wall synthesis, nucleic acid synthesis (DNA replication and RNA transcription), protein synthesis (translation), the synthesis of essential metabolites, and maintaining membrane integrity.

The cell wall is an essential component of bacteria that provides shape and rigidity, and it also serves as a protective barrier against osmotic pressure. Almost all bacteria cell walls are composed of peptidoglycan, a complex polysaccharide that contains N-acetylglucosamine (NAG) and N-acetylmuramic acid (NAM) cross-linked by peptides. This structural feature of bacteria is not present in animals and plants, so inhibiting the synthesis of peptidoglycan selectively targets only bacterial cells. Penicillins are a class of antibiotics that inhibit the formation of the peptide cross-linker between the NAG-NAM polysaccharides in the peptidoglycan cell wall. When Gram-positive bacteria are exposed to penicillin, the drug inhibits peptidoglycan cell wall synthesis, causing bacterial cells to become weak and eventually lyse. However, some bacteria, such as *Staphylococcus* species, have developed resistance to penicillins by producing penicillin-cleaving enzymes called penicillinases. As penicillins were used more and more, the penicillinase genes quickly spread among bacteria to survive, making natural penicillins useless in many infections. To overcome these limitations of natural penicillins, new antibiotics resistant to penicillinases have been developed. These antibiotics inhibit peptidoglycan cell wall synthesis and are resistant to penicillinases. Examples of these antibiotics include cephalosporins, oxacillin, methicillin, vancomycin, carbapenems, and monobactams.

One of the most common mechanisms by which antibiotics work is by inhibiting the protein synthesis of microorganisms. Proteins are essential components of cellular structures and perform most metabolic activities. Ribosomes are responsible for synthesizing proteins in both prokaryotic and eukaryotic cells. However, eukaryotic ribosomes are bigger and structurally different from prokaryotic ribosomes. Chemicals that inhibit protein synthesis bind to different parts of the ribosome. Antibiotics that only affect bacterial ribosomes, such as chloramphenicol, tetracyclines, erythromycin, aminoglycosides (streptomycin, kanamycin, neomycin, gentamycin, etc.), are used to control bacterial growth.

Nucleic acid synthesis pathways such as replication and transcription are critical metabolic processes in organisms. Some antibiotics target nucleic acid synthesis, but their kinds are not as diverse as those targeting cell wall synthesis or protein synthesis. Rifampin (also called rifampicin), the best-known rifamycin, inhibits mRNA synthesis. It is one of the commonly used antituberculosis antibiotics along with isoniazid because it can penetrate tissues and reach in cerebrospinal fluid and abscesses where *Mycobacterium tuberculosis* can penetrate. Isoniazid, another popular antituberculosis antibiotic, inhibits the synthesis of the waxy lipid, mycolic acids.

Some antibiotics inhibit the packaging of DNA inside bacteria. Prokaryotic chromosomal DNA is usually ~1000 times longer than their cell length when it is stretched, so DNA must have some ordered and compact structure to fit inside microbes. One of the mechanisms to give a compact structure to DNA in bacteria is performed by an enzyme called gyrase. Gyrase introduces supercoils into DNA by breaking and reconnecting DNA strands. Fluoroquinolones inhibit the function of gyrase, so inhibit the growth of bacteria. Ciprofloxacin (commonly called Cipro) and recently developed Gemifloxacin and Moxifloxacin are fluoroquinolones.

Some antibiotics kill bacteria by inhibiting the activity of enzymes in an essential metabolic pathway. Generally, this type of antibiotic has a similar structure to the substrates of the enzymes and competes with the substrates, so they are called a competitive enzyme inhibitor. One of the essential bacterial metabolic pathways in many bacteria is the folic acid synthesis pathway. Sulfonamides, also called sulfa drugs, and trimethoprim are antibiotics inhibiting the folic acid pathway. Since humans do not have the folic acid synthesis pathway, sulfonamides and trimethoprim do not affect human cell growth.

Antibiotic resistance developed by microorganisms is currently a worldwide public health problem. Microbes develop antibiotic resistance through several ways. One way is by blocking the entry of antibiotics into microbes. To enter inside microbial cells, most antibiotics must pass through membrane barriers. Antibiotics are water-soluble, so they cannot penetrate the membranes whose main component is lipids. So, most antibiotics enter inside microbes through transmembrane proteins called porins. If microbes change the structure of a porin protein, the antibiotic that uses the porin to access the microbial cell cytosol cannot enter inside the microbe anymore. Another way is the inactivation of an antibiotic by microbial enzymes. Penicillinase is an example of this method. Penicillinase cleaves the β-lactam ring that confers the inhibition activity of cell wall synthesis to penicillins. Additionally, microbes gain resistance to antibiotics by altering antibiotic target molecules. A lot of antibiotics inhibiting protein synthesis bind to bacterial ribosomes. If bacteria change the structure of antibiotic binding sites on the ribosome without changing the main function of the ribosome, the antibiotic does not work anymore. Lastly, microbes can develop resistance by pumping out the antibiotic inside a microbe by activating existing drug pumps or expressing a new drug pump. Microbes use membrane proteins called drug pumps or efflux pumps to pump out toxic materials inside cells such as antibiotics, detergents, and heavy metals. If microbes change the structure of an existing efflux pump or produce a new pump to expel the antibiotic, the antibiotic cannot be effective in killing the microbes anymore.

Microorganisms employ two routes to obtain the antibiotic resistance mechanisms explained previously. One way is through spontaneous mutations in critical chromosomal genes such as the genes of ribosome subunits, efflux pumps, or porins. These mutations can change the structure of antibiotic targets such as ribosome subunits, efflux pumps, and porins. These mutations are transferred to their offspring through vertical gene transfer. The other way is to acquire new genes from other microbes that are already resistant to antibiotics. The genes or set of genes are usually on plasmids or transposons and can be transferred from one microbe to another through the transfer of plasmids or transposons. This type of gene transfer is called horizontal gene transfer. Microbes can acquire the genes of penicillinases and new efflux pumps by horizontal gene transfer.

To know the amount of an antibiotic necessary for the treatment of an infection, the minimum inhibitory concentration of the antibiotic must be experimentally measured. The minimum inhibitory concentration is the lowest concentration of an antibiotic to successfully control the growth of a microbe. In this exercise, the disk-diffusion method you used to test the effectiveness of disinfectants will be used to test the efficacy of antibiotics against a certain microbe (**Figure 13.1**).

Procedure (individual exercise)

Materials needed

- Liquid cultures of *Escherichia coli, Pseudomonas aeruginosa,* and *Staphylococcus aureus* per lab bench
- Mueller-Hinton agar plates per student (Mueller-Hinton agar is a loose agar that allows for better diffusion of the antibiotics than most other media)
- Antibiotic disks
- One beaker with 95% ethanol per lab bench
- Sterile cotton swabs
- Forceps

The instructor will provide six antibiotic disks. Write their names and concentrations or units below. Popular antibiotic disks used are Ampicillin (10 µg), Bacitracin (10 units), Chloramphenicol (30 µg), Erythromycin (15 µg), Gentamicin (10 µg), Penicillin G (10 units), Streptomycin (10 µg), Tetracycline (30 µg), and Vancomycin (30 µg).

Number	Antibiotic name	Concentration or unit
1		
2		
3		
4		
5		
6		

Day 1

1. Each student will use one bacterium. Choose one among the bacteria provided.
2. Using a sterile cotton swab, spread the bacterium evenly on the Mueller-Hinton plate to make a lawn of bacteria. This can be achieved by swabbing the entire surface of the plate three times, rotating the plate 90° first and then 45° the second time. Discard the swabs in a biohazard bag after using them.
3. Label your plate with the initials of your name and the bacteria you plated.
4. Mark 1 to 6 on the bottom of the plate like **Figure 13.1 in the previous exercise**.
5. Flame-sterilize the forceps as you did **in the previous exercise**.
6. Cool down the forceps in the air and pick up an antibiotic disk from one of the glass petri dishes.
7. Place it on the agar surface marked with a number (if the lab has disk dispensers, your instructor will show you how to use the disk dispensers). Each disk has letters and a number indicating the antibiotic name and amount (e.g., P10 is an abbreviation for penicillin 10 units).
8. Tap it down gently with the forceps to ensure that it adheres to the surface of the agar.
9. Repeat steps 5–8 for the antibiotic disks to be tested.
10. Incubate the plates for 24–48 hours, and store at 4°C until the next lab period.

Day 2

1. Observe your Mueller-Hinton plate and measure zones of inhibition of antibiotics by measuring the diameters of clear zones. Also, observe the Mueller-Hinton plates of your peers, so you can fill out **Table 14.1**. Use **Table 14.2** for the interpretation of your results.

TABLE 14.1 The size of the zone of inhibition. (Each student should fill out their results in **Table 14.1.A** that the instructor attached on a white board.)

Bacteria	# on the plate	Antibiotic name and concentration	Zone of inhibition (mm)	Interpretation*
E. coli				
	1			
	2			
	3			
	4			
	5			
	6			
S. aureus				
	1			
	2			
	3			
	4			
	5			
	6			
P. aeruginosa				
	1			
	2			
	3			
	4			
	5			
	6			

*Resistant, Intermediate, or Sensitive

TABLE 14.1.A There will be 18 tables attached to a white board. Each student should fill out six tables.

Bacterium name:

 # on the plate:

Antibiotic name and concentration:

Initials of students	Zone of inhibition (mm)
Average value	

TABLE 14.2 Zone diameter interpretive standards chart for the determination of antibiotic sensitivity and resistance status by the disk diffusion method.

Antibiotics	Antibiotic amount per disk	Diameter of zone of inhibition (mm)		
		Sensitive	**Intermediate**	**Resistant**
Ampicillin	10 µg	≥17	14–16	≤13
Bacitracin	10 units	≥13	9–12	≤8
Chloramphenicol	30 µg	≥18	13–17	≤12
Erythromycin	15 µg	≥23	14–22	≤13
Gentamicin	10 µg	≥15	13–14	≤12
Neomycin	30 µg	≥17	13–16	≤12
Penicillin G	10 units	≥15	–	≤14
Streptomycin	10 µg	≥15	12–14	≤11
Tetracycline	30 µg	≥19	15–18	≤14
Vancomycin	30 µg	≥17	15–16	≤14

Open Microbiology Journal, 11(1), March 2017, DOI:10.2174/1874285801711010031

Review questions

1. Why is the Mueller-Hinton agar used for an antibiotic sensitivity test but not for a disinfectant sensitivity test?

2. Fill in the blanks in the table below.

Antibiotic	Mechanism of action*	Spectrum of activity%
Ampicillin	Inhibition of	
Bacitracin	Inhibition of	
Chloramphenicol	Inhibition of	
Erythromycin	Inhibition of	
Gentamicin	Inhibition of	
Neomycin	Inhibition of	
Penicillin G	Inhibition of	
Streptomycin	Inhibition of	
Tetracycline	Inhibition of	
Vancomycin	Inhibition of	

*Inhibition of cell wall synthesis, DNA synthesis, RNA synthesis, or protein synthesis.
%Broad spectrum or narrow spectrum. If it is narrow spectrum, it is only effective to Gram-positive or Gram-negative bacteria.

3. Explain what the zone of inhibition is.

4. How do microorganisms develop resistance to antibiotics?

5. Do antibiotics show the same killing effect to different microorganisms? Describe your observation.

EXERCISE 15

STREPTOCOCCAL PATHOGENS

LEARNING OBJECTIVES

After this exercise, you should be able to

1. Know why *Streptococcus* species are catalase-negative.
2. List the diseases caused by each streptococcal pathogen.
3. Perform the catalase test.
4. Distinguish the three types of hemolysis.
5. Explain how *Streptococcus mutans* initiates dental caries.

Introduction

Streptococci are Gram-positive and form chains of variable length of cells with a round shape. Most of them are aerotolerant and do fermentation to produce ATPs. The final fermentation products are small organic acids such as lactic acid and acetic acid. Streptococci in the host's upper respiratory tract usually require 5% to 10% CO_2 for optimal growth.

Even though streptococci do not use oxygen to produce ATP, they can tolerate harmful oxygen radicals. While many facultative aerobes use catalase to eliminate the hydroxyl radical, one of the toxic oxygen species, streptococci do not rely on catalase for this purpose. Instead, they use peroxidase to remove the hydroxyl radical. Catalase transforms hydrogen peroxide (H_2O_2) into water (H_2O) and oxygen

TABLE 15.1 Diseases caused by pathogenic streptococci.

Name	Normal habitat	Diseases
Streptococcus pyogenes	Upper respiratory tract and skin	Pharyngitis (strep throat), impetigo, rheumatic heart disease (endocarditis), rheumatic fever, scarlet fever, glomerulonephritis
Streptococcus agalactiae	Intestinal tract and female genital tract	Bacteremia, puerperal sepsis, meningitis, endocarditis
Streptococcus pneumoniae	Upper respiratory tract	Pneumonia
Streptococcus mutans	Mouth	Tooth decay

gas (O_2), which produces oxygen bubbles upon reacting with hydrogen peroxide. In contrast, peroxidase does not produce oxygen bubbles when it reacts with hydrogen peroxide. As streptococci are catalase-negative, they do not generate oxygen bubbles when mixed with hydrogen peroxide.

Various streptococcal pathogens exist, including *Streptococcus pyogenes, Streptococcus pneumoniae, Streptococcus agalactiae*, and *Streptococcus mutans*. **Table 15.1** presents a list of diseases caused by these streptococcal pathogens.

A significant number of streptococcal pathogens are known to produce hemolysins that result in hemolysis or the degradation of red blood cells on blood agar. Hemolysis is typically categorized into three types: alpha, beta, and gamma (**Figure 15.1**). Identification of these streptococcal pathogens can be presumptively made based on their hemolytic properties.

FIGURE 15.1 Hemolysis types on a 5% sheep blood agar plate.

- Alpha-hemolysis: Partial lysis of red blood cells. Red blood cells are partially destroyed. A greenish opaque zone can appear around the colonies. The greenish color is due to the reduction of hemoglobin.
- Beta-hemolysis: Complete lysis of red blood cells. Lysed red blood cells do not show color, so a clear zone appears around the colonies.
- Gamma-hemolysis: No sign of lysis of red blood cells.

Certain hemolysins are incapable of functioning in the presence of oxygen, making them difficult to observe from surface colonies. Microbes can be inoculated into the agar by piercing it with an inoculation loop to create a low-oxygen environment, allowing the microbes to grow inside the agar. When a microbe produces an oxygen-sensitive hemolysin, in addition to an oxygen-insensitive hemolysin, a larger clear zone surrounding the stabbed area will be visible.

The CAMP test is a laboratory technique that utilizes the presence of the CAMP factor protein to identify specific species of streptococcal bacteria. The CAMP factor enhances the hemolytic activity of beta-hemolytic *Staphylococcus aureus*, leading to a larger zone of hemolysis (**Figure 15.2**). This simple, cost-effective method is particularly valuable for identifying *Streptococcus agalactiae* (group B streptococcus) in clinical samples. Its use is particularly relevant in the diagnosis of infections among pregnant women and newborns, as well as in the prevention of neonatal sepsis.

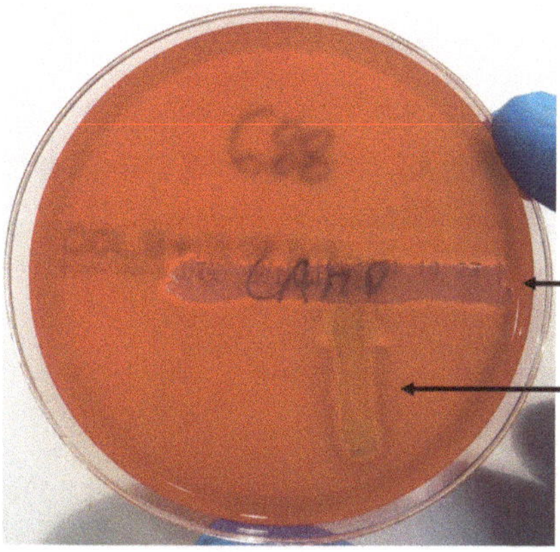

FIGURE 15.2 The result of a positive CAMP test. The arrowhead-type hemolysis shows the sign of a positive CAMP test.

The laboratory technique known as the CAMP test involves observing the interaction between a test organism and a test strain of *Staphylococcus aureus* on an agar plate. *Staphylococcus aureus* is streaked on one side of the plate, while the test organism is streaked perpendicularly. After incubation, if the test organism produces the CAMP factor, an area of augmented hemolysis becomes visible at the intersection of the two organisms.

Streptococcus mutans plays a crucial role in developing dental caries or tooth decay. This bacterium can produce large quantities of dextran, polymers of glucose molecules, that adhere to the tooth surface. Using this dextran, *S. mutans* forms biofilms that shield the bacteria from the host's immune system and make it difficult to remove the bacteria from the tooth surface. The process of dextran formation by *S. mutans* involves the enzyme glucosyltransferase, which splits sucrose molecules and forms dextran from the glucose units. The ability of *S. mutans* to produce dextran and form biofilms on tooth surfaces under sucrose-rich conditions is considered a significant factor in the development of tooth decay. *S. mutans* is one of the major cariogenic organisms that can produce large quantities of acid, which leads to the demineralization of tooth enamel and the formation of cavities.

Materials needed

- Culture plates of *Streptococcus pyogenes*, *Streptococcus agalactiae*, *Streptococcus mutans*, *Streptococcus pneumoniae,* and *Staphylococcus aureus* per lab bench (Streptococcal species should be grown in candle jars)
- Hydrogen peroxide
- Gram-staining reagents
- One slide glass per student
- One sheep blood agar plate per student
- One TSA per student
- One TSA containing 5% sucrose per pair of students

Procedure 1: Catalase test (individual exercise)

1. Pick one streptococcus species.
2. Perform Gram staining and observe them under a microscope using the 100X oil objective lenses.
3. Share your results with your lab mates and fill in the blanks in **Table 15.2** (each student should fill out their results in **Table 15.2A**, which the instructor attached on a whiteboard).
4. Perform the catalase test using the *Streptococcus* species you chose and *Staphylococcus aureus* as a positive control. Follow the procedure below to perform the catalase test. Use one side of a slide glass to test a microbe so that you can test both *Streptococcus* species and *Staphylococcus aureus* on a slide glass.
    a. Place a drop of hydrogen peroxide on a glass slide.
    b. Take a small amount of bacterial culture with the flame-sterilized inoculating loop and mix it well with a hydrogen peroxide drop.
    c. Catalase-positive microbes will show bubbles.
5. Share your results with your lab mates and fill in the blanks in **Table 15.2** (each student should fill out their results in **Table 15.2.A**, which the instructor attached on a whiteboard).

TABLE 15.2 Fill in the blanks in the table based on your peers' and your results.

	Gram type (+ or –) and cell shape[1]	Catalase test (+ or –)
Streptococcus pyogenes		
Streptococcus pneumoniae		
Streptococcus agalactiae		
Streptococcus mutans		
Staphylococcus aureus		

[1] Round, rod, spiral, string, cluster, etc.

Procedure 2: Antibiotic resistance of streptococci (individual exercise)

Day 1

1. Do streak-plating using an inoculation loop for isolating colonies on a sheep blood agar plate with the *Streptococcus* species you chose. The first streak needs heavy inoculum to make a bacterial lawn on the first streak area.
2. Sterilize forceps by dipping the tip of forceps into ethanol, flaming it, and waiting until the flame goes out.
3. Pick up one disk of optochin or bacitracin with the forceps and place it on the first streak zone on the streak plate where the inoculum is heaviest (**Figure 15.3**). The instructor will tell you what antibiotic you should use. Tap the disk gently with your forceps so it does not fall off when the plate is turned upside down.
4. With a flame-sterilized inoculating loop, make several stabs into the agar in the heavily inoculated region of the plate, but away for the optochin or bacitracin disk to deposit some cells under the surface of the agar (**Figure 15.3**). Later, the stabbed area will be observed for oxygen-sensitive hemolysis.
5. Incubate the plates upside down in a candle jar (anaerobic jar with a candle) at 37°C for 24–48 hours, then store at 4°C (**Figure 15.4**). In a candle jar, the oxygen level reduces from 21% to 17%, and the CO_2 level increases to between 5% to 10%.

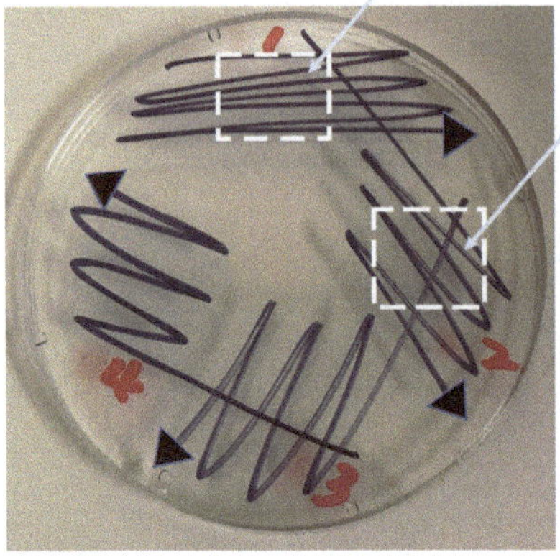

FIGURE 15.3 The locations where an antibiotic disk is placed and several stabs are made.

FIGURE 15.4 A commercial anaerobic jar.

Day 2

1. Observe the sheep blood agar plates for hemolysis. Is hemolysis present around isolated surface colonies? Around the stabs? What kind of hemolysis is present? Record these results in **Table 15.3**. (Each student should fill out their results in **Table 15.3A**, which the instructor attached on a whiteboard.)
2. Observe the sheep blood agar plates for the growth or inhibition of growth around the optochin or bacitracin disk. Measure and record the zone of inhibition from the edge of the disk out to where growth occurred in **Table 15.3**.

TABLE 15.3 Hemolysis type and antibiotic sensitivity of streptococcal pathogens.

	Hemolysis location and degree[1]	Types of hemolysis[2]	Optochin inhibition[3]	Bacitracin inhibition[4]	The CAMP test (+ or −)
Streptococcus pyogenes					Not determined
Streptococcus pneumoniae					
Streptococcus agalactiae					
Streptococcus mutans					

[1] The location where hemolysis occurs (around isolated surface colonies, the stabs, or both). If hemolysis occurred in both places, which clear zone is bigger?
[2] What type of hemolysis (α, β, or γ) does the bacterium show?
[3] Sensitive or resistant based on the presence of a clear zone caused by optochin.
[4] Sensitive or resistant based on the presence of a clear zone caused by bacitracin.

Procedure 3: CAMP test (individual exercise)

Day 1

1. Make a streak of *S. aureus* across the top of a sheep blood agar plate (**Figure 15.5**).
2. Write the initials of *Streptococci* on the bottom of the blood agar plate and streak each of the *Streptococci* (*Streptococcus pneumoniae, Streptococcus agalactiae*, *Streptococcus mutans*) in a straight line perpendicular to, but not touching, the *S. aureus* streak (**Figure 15.5**).
3. Incubate the plates upside down in candle jars at 37°C for 24–48 hours, and then store at 4°C.

Day 2

1. Observe the CAMP test for an arrowhead-shaped area of hemolysis at the end of a streptococcal streak pointing toward the *S. aureus* streak.
2. Share your results with your peers and fill in the blanks in **Table 15.3**. (Each student should fill out their results in **Table 15.3.A** that the instructor attached on a whiteboard.)

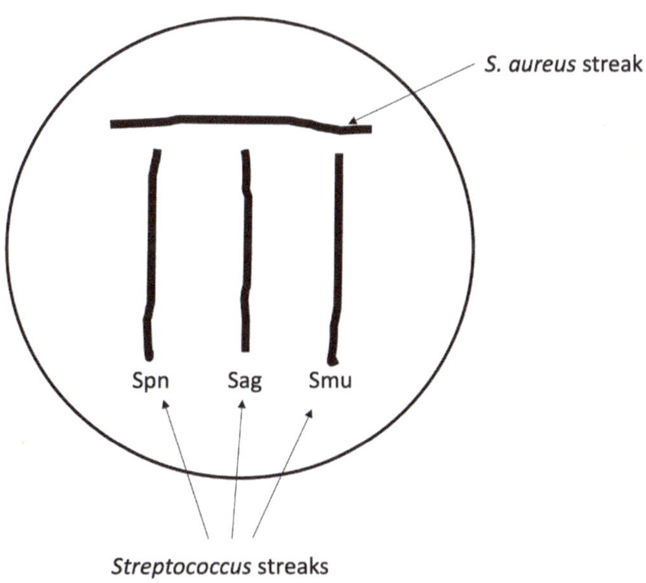

FIGURE 15.5 CAMP test.

Procedure 4: Extracellular polysaccharide formation by *Streptococcus mutans* (individual exercise)

Day 1

1. Perform streak plating on a TSA and a TSA containing 5% sucrose with *Streptococcus mutans* and incubate in a candle jar at 37°C for 24–48 hours, and then store at 4°C.

Day 2

1. Observe the appearance of the plates. Which plate has mucoid cells?

TABLE 15.2.A Fill in the blanks in the table based on your peers' and your results. The instructor will attach this table to a whiteboard. Each student should fill out after finishing Procedure 1. Based on these results, fill out **Table 15.2**.

Students' initials	Microorganism name	Gram type (+ or –) and cell shape[1]	Catalase test (+ or –)

[1] Round, rod, spiral, string, cluster, etc.

TABLE 15.3.A Hemolysis type and antibiotic sensitivity of Streptococcal pathogens. The instructor will attach this table to a whiteboard. Each student should fill out after finishing Procedures 2 and 3. Based on these results, fill out **Table 15.3**.

Initials of students	Micro-organism	Hemolysis location and degree[1]	Types of hemolysis[2]	Optochin inhibition[3]	Bacitracin inhibition[4]	The CAMP test (+ or −)

[1] The location where hemolysis occurs (around isolated surface colonies, the stabs, or both). If hemolysis occurred in both places, which clear zone is bigger?
[2] What type of hemolysis (α, β, or γ) does the bacterium show?
[3] Zone of inhibition (mm) caused by optochin.
[4] Zone of inhibition (mm) caused by bacitracin.

Review questions

1. What are the meanings of strep and coccus? Do the meanings match your observation?

2. Many *Streptococci* are medically important pathogens. What diseases are caused by each Streptococcal species?

3. What does catalase positive mean?

4. If hemolysis occurs only around stabs, not isolated surface colonies, or vice versa, what do these mean?

5. Which *Streptococci* are sensitive to optochin or bacitracin?

6. Which streptococcal species were CAMP test-positive in this exercise?

7. Why are candle jars used to grow *Streptococci*?

8. What characteristic of *Streptococcus mutans* contributes to cause dental caries?

Credits

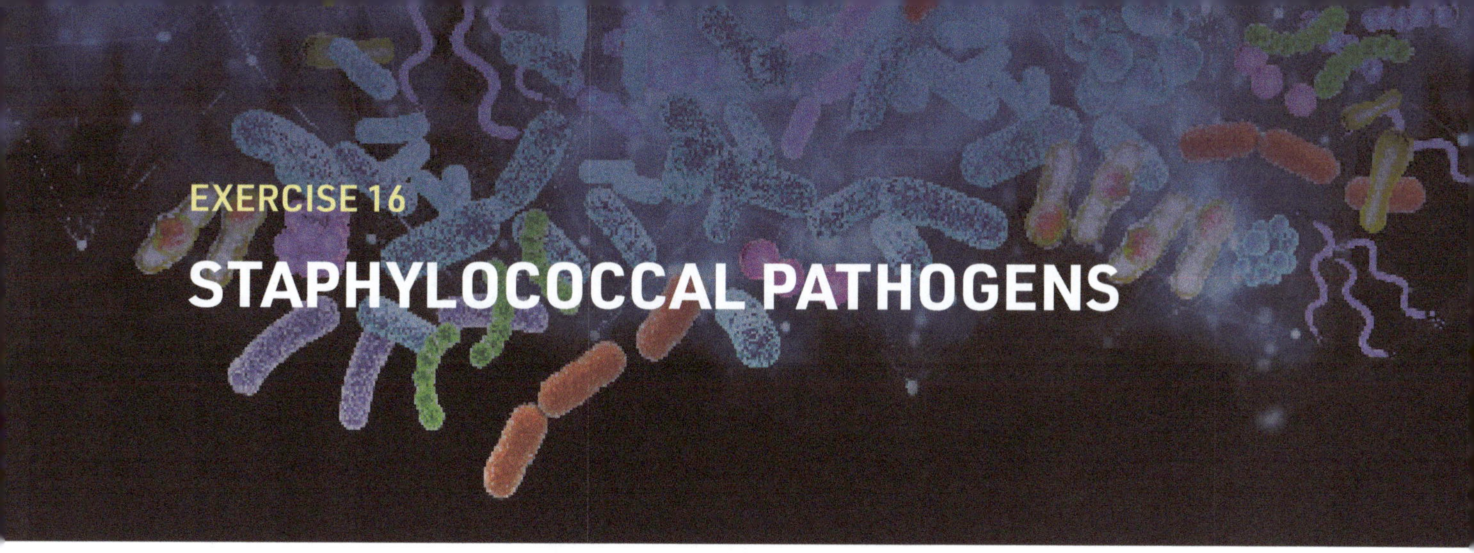

EXERCISE 16

STAPHYLOCOCCAL PATHOGENS

LEARNING OBJECTIVES

After this exercise, you should be able to

1. Explain the characteristics of *Staphylococcus*.
2. Define opportunistic pathogens.
3. Explain how the MSA plates work to distinguish *Staphylococcus aureus* from other microbes.
4. Know why some pathogens form blood clots around them at the initial stage of infection.

Introduction

Staphylococcus species are Gram-positive bacteria and common normal flora. They are found on the skin and mucosal membranes and, usually, they do not cause harm. However, they can become opportunistic pathogens when they enter human tissues. Unlike *Streptococcus*, *Staphylococcus* species form clusters, which can help distinguish them from other bacterial types.

This exercise will use two *Staphylococcus* species: *Staphylococcus aureus* and *Staphylococcus epidermidis*. *Staphylococcus aureus* is an important opportunistic pathogen. *Staphylococcus aureus* can cause food poisoning and a variety of diseases, such as skin infections, endocarditis, and toxic shock syndrome. *Staphylococcus epidermidis* are usually not pathogenic. The mannitol salt agar (MSA) plates are used in this exercise to distinguish these microbes. MSA plates contain

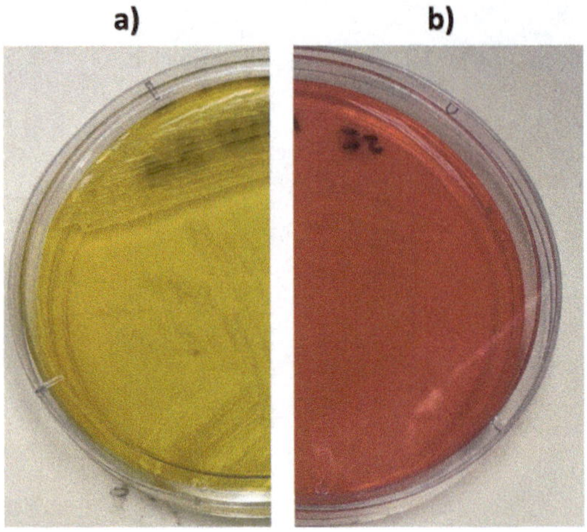

FIGURE 16.1 MSA plates with the growth of *Staphylococcus aureus* (a, yellow) or *Staphylococcus epidermidis* (b, red).

7.5 sodium chloride, creating a hypertonic environment inhibiting the growth of most bacteria, including *Streptococcus*. However, the two *Staphylococcus* species are halotolerant, so they can grow in this high-salt environment. The MSA plates also contain the sugar alcohol mannitol as a carbon source. Only some microorganisms, such as *Staphylococcus aureus,* can ferment mannitol and produce acids as by-products. However, *Staphylococcus epidermidis* cannot ferment mannitol. Instead, *Staphylococcus epidermidis* uses other medium components in MSA plates, peptone, and beef extract as carbon and energy sources, and it does not produce acids. MSA plates also contain phenol red as a pH indicator. When acids are produced by microbes, phenol red changes color from red to yellow. So, only *Staphylococcus aureus* changes the MSA plates to yellow (**Figure 16.1**).

Some microbial pathogens form blood clots to protect themselves at the initial stage of infection. Coagulase is a bacterial enzyme triggering blood clot formation. Coagulase cleaves fibrinogen from the plasma and produces fibrin, which is the main component in blood clots. When pathogens invade a host tissue, nearby blood vessels leak to provide the blood plasma containing immune factors such as white blood cells, antimicrobial peptides, and complement factors to destroy pathogens. Some pathogens, such as *Staphylococcus aureus,* form blood clots around them to protect from the attack by the host immune system.

Materials needed

- Streak culture plates of *Staphylococcus aureus*, *Staphylococcus epidermidis*, and *Streptococcus mutans* (1 each per lab bench)
- One sheep blood agar plate per student

- One mannitol salt agar (MSA) plate per student
- One tube containing 0.1 ml of rabbit plasma per student

Procedure 1: Test of staphylococcal species with differential media (individual exercise)

Day 1

1. Pick one organism among the three organisms provided.
2. Label one MSA plate and one sheep blood agar plate (BAP) with the micro-organism's initials and your initials.
3. Use a sterile inoculating loop and streak for isolation of each organism on a mannitol salt and a sheep BAP. On the sheep BAP, stab the loop into the agar right after you have finished each streak but before flaming the loop.
4. Incubate the plates for 24–48 hours at 37°C and store at 4°C.
5. Perform Gram-staining with the microorganism. Fill the first column of **Table 16.1** with the Gram-staining results of yours and your peers.

Day 2

1. Share the results of your MSA and BAP plates with your peers and record them in **Table 16.1** (each student should fill out their results in **Table 16.1A,** which the instructor attached on a whiteboard).

TABLE 16.1 The characteristics of *Staphylococcus* and *Streptococcus* species.

Microorganism	Gram type[1]	Cell shape[2]	Growth on MSA[3]	Color change on MSA[4]	Hemolysis on BAP[5]	Coagulase test[6]
Staphylococcus aureus						
Staphylococcus epidermidis						
Streptococcus mutans						

[1] Positive or negative
[2] Round, rod, spiral, string, cluster, etc.
[3] Positive or negative
[4] Original color to current color
[5] α, β, or γ
[6] Positive or negative

Procedure 2: Coagulase test (individual exercise)

Day 1

1. Get a tube containing 0.1 ml of rabbit plasma from the instructor. Label the tubes with your initials and the microorganism you picked: *Staphylococcus epidermidis*, *Staphylococcus aureus*, or *Streptococcus mutans*.
2. Inoculate the tubes with the microbe you picked.
3. Incubate the tube for more than three hours at 37°C and store at 4°C.

Day 2

1. Record which microbe coagulates the rabbit plasma in **Table 16.1** (each student should fill out their results in **Table 16.1.A**, which the instructor attached on a whiteboard).

Review questions

1. Fill in the blanks in the table below. What happens when the bacteria below are inoculated on the MSA plates?

Components in MSA→ Bacteria ↓	7.5% salts (growth or no growth)	Mannitol (fermentable or not fermentable)	Phenol red (red or yellow)
Staphylococcus aureus			
Staphylococcus epidermidis			
Streptococcus mutans			

2. What is the benefit when pathogens make blood clots around them during infection?

3. What is the definition of opportunistic pathogens?

TABLE 16.1.A The characteristics of *Staphylococcus* and *Streptococcus* species. The instructor will attach this table on a whiteboard. Each student should fill out this table after finishing Procedures 1 and 2. Based on these results, fill out **Table 16.1**.

Microorganism	Gram type[1]	Cell shape[2]	Growth on MSA[3]	Color change on MSA[4]	Hemolysis on BAP[5]	Coagulase test[6]

[1] Positive or negative
[2] Round, rod, spiral, string, cluster, etc.
[3] Positive or negative
[4] Original color to current color
[5] α, β, or γ
[6] Positive or negative

EXERCISE 17

INTESTINAL PATHOGENS OF *ENTEROBACTERIACEAE*

LEARNING OBJECTIVES

After this exercise, you should be able to

1. Explain differential media and selective media, and when they are used.
2. List the differential media for lactose-fermenting *Enterobacteriaceae* and explain how they work.
3. Know how to use the Enterotube system.
4. Interpret the result of the Enterotube system to identify *Enterobacteriaceae*.

Introduction

Enterobacteriaceae is a family of Gram-negative bacilli that are aerobes or facultative anaerobes. Their ability to produce enzymes that can break down complex molecules, such as sugars and proteins, allows them to survive and grow in a variety of environments, including soil, water, plants, and the gastrointestinal tract of humans and animals. The family includes more than 35 genera and over 100 species, including some genera called enteric bacteria. Some of the most well-known members of the *Enterobacteriaceae* family include *Escherichia coli*, *Proteus*, *Salmonella*, *Klebsiella*, *Shigella*, and *Yersinia*, many of which can cause disease in humans and animals. These bacteria can cause a range of illnesses, including foodborne illnesses, urinary tract infections, pneumonia, and sepsis (**Table 17.1**). Treatment of

Enterobacteriaceae infections often involves the use of antibiotics, although antibiotic-resistant strains are becoming more common and pose a significant public health threat. *Enterobacteriaceae* are often identified using a variety of laboratory tests, including the Gram stain, biochemical tests, and DNA sequencing.

One of the ways to presumptively identify and isolate microorganisms is using selective or differential media. Selective media are designed to inhibit the growth of certain types of microorganisms while allowing others to grow. These media contain compounds that selectively inhibit the growth of some microorganisms while allowing others to grow. For example, MacConkey agar is a selective medium that is commonly used to isolate Gram-negative bacteria while inhibiting the growth of Gram-positive bacteria. This selectivity is achieved by incorporating bile salts and crystal violet into the medium. Differential media, on the other hand, are designed to differentiate between different types of microorganisms based on their metabolic characteristics. These media contain compounds that allow certain types of microorganisms to grow and produce visible differences in the appearance of the colonies. For example, blood agar is a differential medium that is commonly used to differentiate between different species of *Streptococcus* based on their ability to produce hemolysins. MacConkey agar is also a differential medium. Lactose-fermenting bacteria such as *Escherichia coli* and *Klebsiella* will ferment lactose and produce acid, causing the neutral red indicator to turn pink or red. Nonlactose fermenting bacteria such as *Salmonella* and *Shigella* will not produce acid and remain colorless. Thus, MacConkey agar is a selective and differential medium. MSA (mannitol salt agar) you used in the exercise of *Staphylococcus* is another example of a medium that is both selective and differential. Several selective and differential media can be used to presumptively identify or isolate the cells of *Enterobacteriaceae* (**Table 17.2**).

TABLE 17.1 Diseases caused by pathogenic *Enterobacteriaceae*.

Enterobacteriaceae	**Diseases caused**
Escherichia coli	Diarrhea, urinary tract infections, kidney failure, etc.
Klebsiella	Urinary tract infections, pneumonia, bloodstream infections, wound infections, etc.
Salmonella	Salmonellosis (gastroenteritis with diarrhea, abdominal pain, fever, and vomiting), typhoid fever
Shigella	Shigellosis (diarrhea, often dysentery [bloody diarrhea], fever, and abdominal pain)
Yersinia	Plague, diarrhea
Proteus	Urinary tract infections

TABLE 17.2 Lactose fermentation detection agar media (**Figure 17.1**): All of them are differential and selective media (bile salts, and some dyes such as Eosin Y, methylene blue, and crystal violet inhibit the growth of many Gram-positive bacteria).

Media (agar plates)	Colony color of nonlactose fermenters	Colony color of lactose fermenters	pH indicators or dyes in media
EMB (Eosin methylene blue)	Colorless	Green metallic sheen	Eosin Y, methylene blue
MAC (MacConkey)	Colorless	Magenta	Neutral red, crystal violet
HEA (Hektoen enteric agar)	Green to blue	Salmon-pink to orange	Acid fuchsin, bromothymol blue

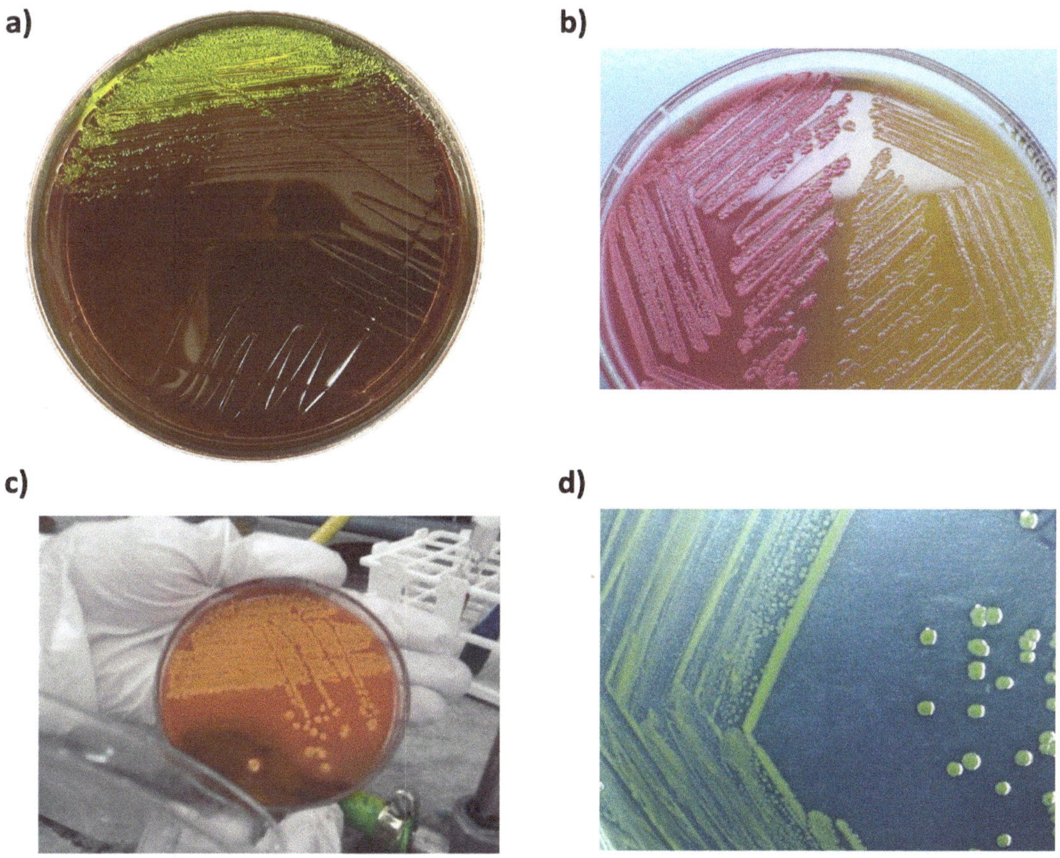

FIGURE 17.1 EMB, MAC, and HEA agar plates: a) *E. coli* grown on EMB; b) lactose-fermenting (left) and non fermenting (right) bacteria on MAC; c) lactose fermenting bacteria on HEA; d) lactose-non fermenting bacteria on HEA.

Enterotube is a plastic tube that contains 12 different media designed to identify enteric bacteria (**Figure 17.2**). This system provides a rapid and reliable method for identification of *Enterobacteriaceae*. All tests are done simultaneously by inoculation from a single isolated colony. The Enterotube test is based on 15 different biochemical reactions, including glucose fermentation, gas production from glucose, lysine decarboxylase, ornithine decarboxylase, hydrogen sulfide production, indole production, adonitol fermentation, lactose fermentation, arabinose fermentation, sorbitol fermentation, Voges-Proskauer reaction, dulcitol fermentation, and urease production. The Enterotube can be used to differentiate between enteric pathogens such as *Salmonella* and *Shigella*, as well as other enteric bacteria. However, it should be noted that the test is presumptive, and additional confirmatory tests may be required for accurate identification. Overall, the Enterotube system is a convenient and efficient method for the preliminary identification of enteric bacteria.

FIGURE 17.2 Enterotube.

Materials needed

- Culture plates of *E. coli, Salmonella typhimurium, Shigella flexneri*, and *Proteus vulgaris* (each per lab bench)
- 1 EMB, 1 HEA, and 1 MAC agar plate per student
- 1 Enterotube per group
- 1 Enterotube chart per group
- 1 Enterotube scoring sheet and interpretative table per group
- Kovacs' reagent

Procedure 1: Examination of *Enterobacteriaceae* with EMB, HEA, and MAC media (individual exercise)

Day 1

1. Each student will be assigned to use one microorganism.
2. Record the colors of the agar plates provided before inoculating them in **Table 17.3**.
3. Streak the bacterium for isolation on EMB, HEA, and MAC plates with the microorganism provided. After streaking, label your plates with your initials and microorganism's initials.
4. Incubate the plates at 37°C for 24–48 hours and store at 4°C.

Day 2

1. Observe the plates and share your results with your peers by filling out **Table 17.3.A,** attached on a white board. Based on the results in **Table 17.3.A**, fill out the blanks in **Table 17.3**.

TABLE 17.3 Color changes of plates before and after incubation with *Enterobacteriaceae*.

	EMB plate (color of colonies)		HEA plates (color of colonies)		MAC plates (color of colonies)	
Bacteria	**Before**	**After**	**Before**	**After**	**Before**	**After**
E. coli						
Salmonella typhimurium						
Shigella dysenteriae						
Proteus vulgaris						

Procedure 2: Identification of *Enterobacteriaceae* with Enterotubes (group exercise)

Day 1

Inoculate the Enterotube with your microorganism by the following procedure in the manual of the BBL Enterotube II (the instructor will demonstrate this exercise).

1. Remove both caps of the Enterotube. The tip of the inoculating wire is under the white cap. Do not flame the wire.
2. Pick a well-isolated colony directly with the tip of the Enterotube inoculating wire. A visible amount of inoculum should be seen at the tip and the side of the wire. Avoid touching agar with wire.
3. Inoculate the Enterotube by first twisting the wire, then withdrawing the wire through all twelve compartments applying a turning motion.
4. Reinsert the wire (without sterilizing) into the Enterotube, using a turning motion through all 12 compartments, until the notch on the wire is aligned with the opening of the tube. The tip of the wire should be seen in the citrate compartment. Break the wire at the notch by bending it. The portion of the wire remaining in the tube maintains anaerobic conditions necessary for true fermentation of glucose, production of gas, and decarboxylation of lysine and ornithine.
5. With the broken-off part of the wire, punch holes through the foil covering the air inlets of the last eight compartments (adonitol, lactose, arabinose, sorbitol, Voges-Proskauer, dulcitol/PA, urea, and citrate) to support aerobic growth in these compartments. Replace both caps.
6. Incubate at 37°C for 18–24 hours with the Enterotube lying on its flat surface or in an upright position. Allow for air circulation between incubated tubes.
7. Store at 4°C.

Day 2

1. Interpret and record all reactions except for indole and Voges-Proskauer. All other tests must be read before the indole is performed as the reagents added for these tests may alter the remainder of the Enterotube reactions. Indicate each positive test result by circling the number appearing below the appropriate compartment on the Results Pad.
2. To perform the indole test:
   a. Place the Enterotube horizontally on the bench or vertically in a rack with the glucose compartment pointing downward. With a hot inoculating wire, melt a hole of about 3–4 mm in diameter in the plastic film above the indole/H2S compartments.
   b. Add 3–4 drops of Kovacs' reagent to the H2S/indole compartment. Allow the reagent to contact the surface of the medium. A positive test is indicated by the development of a red color in the added reagent within 10 seconds.
   c. We will **not** perform the Voges-Proskauer test.
3. After performing the indole test, if positive, circle the appropriate numbers on the prepared sheet.
4. Add circled numbers in the bracketed section and enter the sum in the space provided below the arrow.
5. Locate the five-digit number in the interpretation guide (codebook) and find the best answer(s) in the column entitled "ID Value." Make sure to use the right database: Enterobacteriaceae—method without VP.

TABLE 17.3.A Color changes of the plates before and after incubation with *Enterobacteriaceae*.

Microorganism:

Initials of students	EMB plate (color of colonies)		HEA plate (color of colonies)		MAC plate (color of colonies)	
	Before	After	Before	After	Before	After

Review questions

1. What are selective media and differential media and when are they used?

2. Provide examples of selective and differential media and explain why they are selective or differential media.

3. Among the *Enterobacteriaceae* you used, which ones are lactose fermenters?

4. Many enteric pathogens are nonlactose fermenters. Why?

5. What is the name of the microorganism identified by the Enterotube test? Is it the same microorganism used for the Enterotube test?

Credits

Fig. 17.1B: Medimicro, "Lactose-fermenting bacteria (left) and non-lactose-fermenting bacteria (right) on MAC," https://commons.wikimedia.org/wiki/File:MacConkey_agar_with_LF_and_LF_colonies.jpg, 2007.
Fig. 17.1C: Center for Disease Control, "Non-lactose fermenting bacteria," https://commons.wikimedia.org/wiki/File:Gram_negative_Shigella_sonnei_bacteria_which_spent_48_hours_cultured_on_Hektoen_enteric_agar.jpg, 2014.
Fig. 17.1D: Copyright © 2011 by Mfloayza (CC BY-SA 3.0) at https://commons.wikimedia.org/wiki/File:Desarrollo_de_Enterobacterias_fermentadoras_de_Lactosa_en_HEA.jpeg.

EXERCISE 18

HUMAN SAMPLE COLLECTION FOR MICROORGANISM CULTIVATION

LEARNING OBJECTIVES

After this exercise, you should be able to

1. Explain how to collect the blood, urine, stool, nasal, and throat samples from humans.
2. Collect human samples.
3. Interpret the results of human sample analyses.

Introduction

When collecting human samples for microorganism detection, it is important to follow appropriate protocols to ensure the accuracy and reliability of the results. Here are some essential things to consider:

1. Sample type: The type of sample collected will depend on the microorganisms being tested for and the part of the body being sampled. Common sample types include urine, feces, blood, saliva, sputum, and swabs from various body surfaces.
2. Sampling technique: Proper sampling technique is crucial to avoid contamination and accurately represent the microorganisms present. The method will depend on the sample type and can include using sterile collection containers, following aseptic techniques, and collecting multiple samples to increase the chances of detection.

3. Timing: The timing of the sample collection is also important, as some microorganisms may only be present during certain stages of infection or at specific times of day. In some cases, it may be necessary to collect multiple samples over time to increase the chances of detection.
4. Storage and transport: Proper storage and transportation of the samples are crucial to ensure that the microorganisms remain viable and do not become contaminated during transit. Samples should be stored at the appropriate temperature and transported as quickly as possible to the testing facility.
5. Documentation: Documentation of the sample collection process is important to ensure that the results are accurate and can be traced back to the original sample. This includes labeling the samples properly, documenting the time and date of collection, and noting any relevant patient information or symptoms.

Collecting a urine sample is a common medical procedure used to diagnose various conditions, such as urinary tract infections (UTIs), kidney problems, and diabetes. To perform the test, a healthcare provider will ask the person to provide a clean-catch urine sample. This involves cleaning the genital area and collecting a midstream urine sample in a sterile container.

1. Wash your hands thoroughly with soap and water before collecting the urine sample.
2. Choose a clean, sterile container with a tight-fitting lid. You can usually get one from your doctor's office or a pharmacy.
3. Clean your genital area with a mild soap and water. It's important to clean from front to back to prevent contamination with bacteria from the anus.
4. Start urinating in the toilet, and then hold the container under your stream to collect about an ounce (30 milliliters) of urine. Avoid touching the inside of the container or stopping and starting the stream.
5. Finish urinating into the toilet.
6. Tightly close the container with the lid.
7. Label the container with your name, the date and time of collection, and any other relevant information your doctor may require.
8. Bring the urine sample to your doctor's office or the laboratory immediately. The urine is then sent to a laboratory for analysis.

A urine culture test is a laboratory test that checks for the presence of bacteria or other microorganisms in a urine sample. In the laboratory, the urine sample is placed on a special agar plate containing nutrients to help bacteria grow. The plate is then incubated for a certain period, usually 24-48 hours. If there are any bacteria in the urine sample, they will grow on the agar plate and form colonies. The number of colonies will then be counted, and the type of bacteria present will be identified. The interpretation of a urine culture test depends on the number of bacteria present in the sample. A positive culture indicates the presence of bacteria in the urine and is usually reported as the number of colony-forming units (CFUs) per milliliter of urine. A high number of CFUs (typically more than 100,000) indicates a significant bacterial infection. However, a lower number of CFUs may still indicate an infection in some cases. The type of bacteria present is also important, as different bacteria may require different types of antibiotics for treatment.

Feces culture, also known as stool culture, is a laboratory test that is used to identify bacteria, viruses, or parasites that may be present in a person's stool. Feces culture is not always necessary for every case of gastrointestinal symptoms and should be ordered by a healthcare professional based on their clinical judgment. Performing a feces culture test involves the following steps:

1. Collection of the sample: The patient will be provided with a container to collect a fresh stool sample. It is crucial to avoid contamination of the sample with urine or toilet paper.
2. Transport: The sample should be transported to the laboratory immediately. If there is a delay, the sample should be refrigerated to prevent the growth of bacteria.
3. Sample processing: In the laboratory, a small amount of the stool sample will be mixed with a culture medium.
4. Incubation: The stool sample's culture medium will be placed in an incubator.
5. Identification of bacteria: After the bacteria have had a chance to grow, they will be identified by visual inspection, chemical tests, or genetic analysis.
6. Sensitivity testing: If bacteria are identified, they will be tested to see which antibiotics are effective against them.

Blood culture tests are typically only ordered when a patient is suspected of having a bloodstream infection. These infections can be serious and potentially life-threatening, so prompt and accurate diagnosis is essential for effective treatment. Performing a blood culture test involves collecting a small volume of blood and then culturing it in a laboratory to detect any bacteria, viruses, or fungi that may be causing an infection. The steps involved in performing a blood culture test are:

1. Preparation: Before performing the blood culture test, the healthcare provider will make sure that the patient's skin is clean and disinfected to prevent any contamination during the procedure. The healthcare provider will wear gloves and use sterile equipment to collect the blood sample.
2. Blood collection: A small amount of blood is collected from a vein in the arm using a needle and syringe. The healthcare provider will collect two to three

blood samples from different sites to increase the likelihood of detecting any pathogens.

3. Blood culture: The collected blood is transferred to a nutrient-rich medium that promotes the growth of any microorganisms that may be present in the blood. The inoculated medium will be placed for up to five to seven days in a special incubator where the temperature, humidity, and other environmental conditions are carefully controlled to promote the growth of microorganisms.
4. Identification: If microorganisms are detected in the blood culture, each microorganism will be identified, and its susceptibility to various antibiotics will be determined.

Nasal or throat swabs are commonly used to diagnose respiratory infections, such as COVID-19, influenza, and strep throat. These tests are used to detect the presence of viral or bacterial pathogens in the respiratory system. Nasal swabs are typically used to collect samples from the nasal cavity, which is located behind the nose. This type of swab is used to collect samples for COVID-19 testing and other respiratory viruses such as influenza and respiratory syncytial virus (RSV). Throat swabs are used to collect samples from the back of the throat. They are commonly used to diagnose strep throat, which is caused by the bacterium *Streptococcus pyogenes*. Throat swabs may also be used to test for other respiratory viruses.

Materials needed

- Two sterile sample containers per student
- Sheep blood, 1 ml per lab bench
- Blood agar plates, 4 per student
- One MSA plate per student
- One rabbit serum tube per student
- Disposable inoculation loops (10 μl)

Procedure 1: Urine culture (individual exercise)

Day 1

1. You will be provided with two sample containers a week before this exercise. Collect urine samples at home.

2. Label the two sterile containers provided CC (clean catch) or DC (dirty catch).
3. Collect your urine in the container with two different methods below. Urine should be filled around half of the container.
    a. Clean catch technique: Thoroughly clean all external surfaces surrounding the urethra. Discard the first stream of urine. Collect the midstream portion in a sterile specimen container. The container should not touch the skin or clothing.
    b. No clean catch (NCC) technique: Collect a urine sample without any special care.
4. Close the lid tightly and clean the outside of the container with tap water.
5. Store the samples in the refrigerator right away.
6. Bring them to the lab.
7. Take 10 µl urine from your CC sample using a disposable inoculation loop.
8. Add the 10 µl urine sample in the middle of a blood agar plate and spread them all over the surface using the inoculation loop.
9. Perform steps 7 and 8 using the NCC sample.
10. Incubate the plates at 37°C for 24–48 hours and store at 4°C.

Day 2

1. If you see colonies, count the number of colonies on each plate. Determine the number of organisms per milliliter of the urine sample and record in **Table 18.1**.
2. If you see colonies, record your observation on colony morphology, microscopic morphology, and Gram-stain characteristic of the most dominant colony type in **Table 18.1**.

TABLE 18.1 Characteristics of the most dominant colony type on urine culture plates.

	Clean Catch Sample	**Dirty Catch Sample**
CFUs/ml		
Gram-type (+ or −) and cell shape[1]		
Colony morphology		

[1] Round, rod, spiral, string, cluster, etc.

Procedure 2: Blood culture (individual exercise)

Day 1

1. Spread 10 µl blood provided (sheep blood) on a blood agar plate by performing steps 7 and 8 in Procedure 1.
2. Incubate the plates at 37°C for 48 to 72 hours, and store at 4°C.

Day 2

1. If you see colonies, count the number of colonies on the plate. Determine the number of organisms per milliliter of the blood sample.
2. If you see colonies on the plate, record your observation on colony morphology, microscopic morphology, and Gram-stain characteristic of the most dominant colony type in **Table 18.2**.

TABLE 18.2 Characteristics of the most dominant colony type on blood agar plates.

	Blood Culture
CFUs/ml	
Gram-type (+ or −) and cell shape[1]	
Catalase test (positive or negative)	
Colony morphology	

[1] Round, rod, spiral, string, cluster, etc.

Procedure 3. Throat swab (each student will swab their throat) (individual exercise)

Day 1

1. Hold the tongue down with a sterile tongue depressor and rotate a sterile swab over the tonsillar area in the back of the throat. Discard the tongue depressor into the biohazard bag.
2. Rotate the swab on a small area near the edge of a sheep blood agar plate. Discard the swab into the biohazard bag.

3. Streak for isolation from the inoculated area onto the rest of the plate. Make a stab into the agar. Flame your loop when finished.
4. Incubate the plate in a candle jar for 48–72 hours at 37°C.

Day 2

1. Examine the plate for the presence of hemolytic colonies. Gram-stain a colony showing hemolysis and a colony that does not show hemolysis. Are any of them *Staphylococcus* species or *Streptococcus* species? Record your results in **Table 18.3**.

TABLE 18.3 Characteristics of the microorganism isolated.

	Hemolytic colony	Nonhemolytic colony
Gram-type (+ or −) and cell shape[1]		
Catalase test (positive or negative)		
Colony morphology		

[1] Round, rod, spiral, string, cluster, etc.

Procedure 4: Nasal swab (each student will swab their nose) (individual exercise)

Day 1

1. Tilt your head back slightly: This will help to open the nasal passage and make it easier to access the inside of the nostril.
2. Insert a sterile swab into the nostril: Gently insert the swab into the nostril and rotate it several times. You want to make sure that the swab comes into contact with the walls of the nasal passage to collect a sufficient sample.
3. Gently rub the swab across the surface of an MSA plate to spread microbes evenly.
4. Label the plate with the initials of your name.

Day 2

1. Observe the MSA plate. Pick a dominant colony type and perform Gram-staining, catalase test, and coagulase test. Record your results in **Table 18.4**.

TABLE 18.4 Characteristics of the microorganism isolated.

Colony morphology and color change on MSA	Gram type (+ or −) and cell shape[1]	Catalase test (positive or negative)	Coagulase test (positive or negative)

[1] Round, rod, spiral, string, cluster, etc.

Review questions

1. In your urine culture, did you see differences between the clean catch sample and the dirty catch sample? If you did not, what do you think was the problem?

2. If microorganisms grew in the blood culture, what type of microorganisms do you think they are? Explain your answer based on your experimental results.

3. What type of microorganisms do you think you isolated from your throat swab? Explain your answer based on your experimental results.

4. What type of microorganisms do you think you isolated from your nasal swab? Explain your answer based on your experimental results.

EXERCISE 19

PROTEIN PRODUCTION AND PURIFICATION THROUGH BIOENGINEERING

LEARNING OBJECTIVES

After this exercise, you should be able to

1. Explain what biotechnology and genetic engineering are.
2. Describe what expression vectors are.
3. Describe what the promoter and the ribosome-binding site are.
4. Tell why the 6X histidine tag is attached to the protein produced in this exercise.
5. Describe the principle of SDS-PAGE (sodium dodecyl sulfate-polyacrylamide gel electrophoresis).

Introduction

Biotechnology involves the use of living organisms or their components to produce valuable products or carry out specific tasks. Protein production using biotechnology is a crucial process with a wide range of important applications in fields such as medicine, agriculture, and industry. Genetic engineering can be used to change protein characteristics or increase protein production yield, which involves introducing new genes or modifying existing genes in organisms to produce desired traits or products. Genetic engineering has been used to produce medicines such as insulin, to create crops that are resistant to pests or herbicides, and to develop bacteria that can break down waste products.

Plasmids are small extrachromosomal circular DNA and are often used in genetic engineering. Plasmids can be modified to carry foreign DNA sequences and introduced into host cells. Plasmids can be used to produce proteins of interest, such as insulin, which is produced by genetically modified bacteria, and to create transgenic organisms, which have been modified to express specific traits or produce specific products.

Protein expression plasmids or vectors are commonly used in molecular biology research to produce specific proteins of interest. These plasmids contain genetic sequences that allow for the controlled expression of a target protein in a host organism, typically a bacterial or mammalian cell. Protein expression plasmids usually consist of several key elements, including a promoter sequence, a ribosome binding site, and a gene encoding the protein of interest. The promoter sequence is a DNA sequence that regulates the expression of the gene by controlling the rate at which the gene is transcribed into messenger RNA (mRNA). The ribosome binding site is a short DNA sequence that facilitates the binding of ribosomes to the mRNA molecule, allowing for the translation of the mRNA into a protein. Protein expression plasmids may also contain additional sequences that enhance the stability, efficiency, or secreted location of proteins. For example, some plasmids may include a signal sequence that directs the protein to be secreted from the host cell, or a sequence that tags the protein with a specific marker for easy detection or purification. The choice of a protein expression plasmid will depend on a number of factors, including the type of host organism being used, the specific protein being expressed, and the desired level of protein expression.

In this exercise, you will use one of the common expression plasmid systems, the pET system (**Figure 19.1**). The pET expression system is a bacterial expression system widely used to produce recombinant proteins. It is known for its high-level expression of recombinant proteins, which makes it a powerful tool in the field of protein engineering and structural biology. The pET28 plasmid system also allows for tight control of gene expression, which is useful for producing large amounts of recombinant proteins. The pET system is based on the use of plasmids containing the T7 RNA polymerase promoter and the lac operator sequence. In this system, a gene of interest is cloned into the downstream of the T7 promoter in the pET plasmid. The plasmid is then transferred into *E. coli* that has the T7 polymerase gene in its chromosome such as *E. coli* BL21. The T7 polymerase is a bacteriophage enzyme that is not normally present in *E. coli*. Expression of the gene of interest pET28 is induced by adding an inducer such as isopropyl-β-D-thiogalactopyranoside (IPTG), which leads to transcription of the gene by the T7 RNA polymerase.

The expression of a protein in pET28 is under the control of the LacI system, which is a widely used method to control gene expression in recombinant DNA technology. The LacI protein is a repressor that can bind to the operator region of the lac operon in *E. coli*. When LacI is bound to the operator region, it prevents RNA

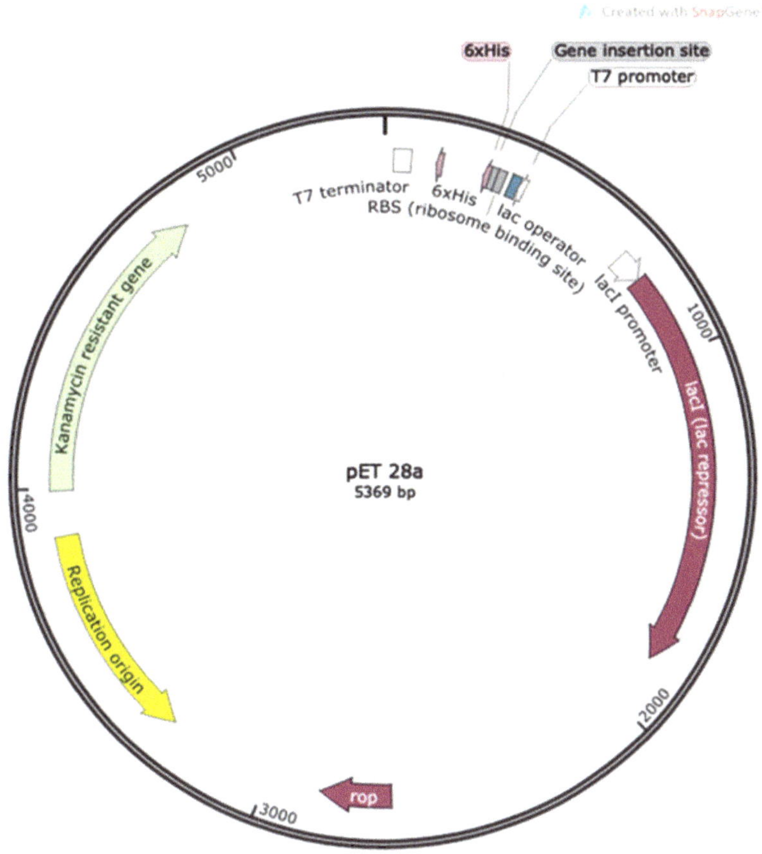

FIGURE 19.1 A map of the pET-28 vector.

polymerase from transcribing the downstream genes of the operon. The pET28 plasmid also contains a lac operator sequence downstream of the T7 promoter. This allows for the addition of IPTG, a lactose analog, which can bind to the LacI protein and cause it to dissociate from the operator region, thereby allowing transcription of the gene of interest.

The 6X histidine-tag (His-tag) purification system is a widely used method for purifying recombinant proteins that have been expressed in bacterial or mammalian cells. The system relies on the specific binding of a polyhistidine (His-tag) sequence, which has been incorporated into the recombinant protein, to a metal ion affinity resin such as a nickel or cobalt resin. The system is a versatile and robust method for purifying recombinant proteins, providing a high yield of purified protein with low levels of contamination. The purification protocol involves several steps. The recombinant protein of interest is first expressed in bacterial or mammalian cells, and the cells are then lysed to release the protein. The lysate is then

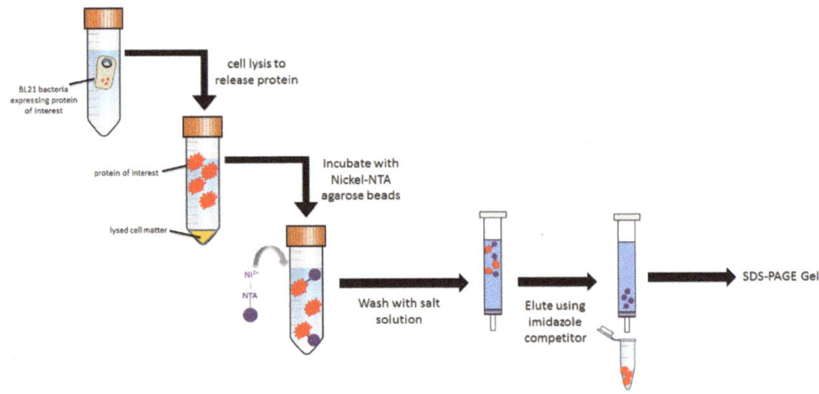

FIGURE 19.2 A purification scheme of a protein tagged with 6X-histidine.

cleared of cellular debris by centrifugation or filtration. The cleared lysate is then incubated with a metal ion affinity resin, such as nickel, or cobalt conjugated agarose, which has a high affinity for the His-tag. The His-tagged protein binds to the resin while other proteins in the lysate are washed away. The resin is then washed with a series of buffers to remove any nonspecifically bound proteins or contaminants. The His-tagged protein is then eluted from the resin using a buffer containing imidazole, which competes with the His-tag for binding to the resin. The protein can also be eluted using other agents such as EDTA, high salt concentration, or changes in pH. The eluted protein is typically buffer exchanged into a suitable buffer for downstream applications and concentrated using a centrifugal filter or other concentration method. The purified protein is analyzed for purity and concentration using various techniques, such as SDS-PAGE, western blotting, and UV spectrophotometry.

Sodium dodecyl sulfate-polyacrylamide gel electrophoresis (SDS-PAGE) is a widely used technique for separating proteins based on their sizes. The principle behind SDS-PAGE is to denature and coat proteins with a negatively charged detergent, sodium dodecyl sulfate (SDS), which causes the proteins to have a uniform negative charge density based on their mass. By running the SDS-coated proteins through a polyacrylamide gel matrix, the negatively charged proteins migrate through the matrix under the influence of an electric field, with smaller proteins moving faster than larger ones.

The gel matrix is typically composed of two parts, the top (stacking) layer and the bottom (resolving or running) layer. The stacking gel has a lower percentage of acrylamide (making a larger pore size) and pH 6.8 (providing a low ionization rate of glycine that is the major buffer component), which help to concentrate the proteins into a tight band before they enter the resolving gel. The resolving gel has a higher percentage of acrylamide (making a smaller pore size) and pH 8.8 (providing a high ionization rate of glycine), allowing for better separation of proteins by their sizes.

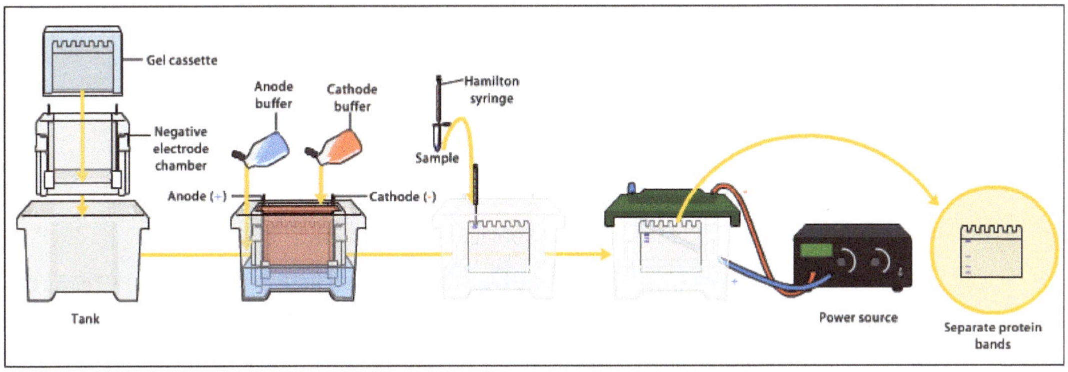

FIGURE 19.3 The procedure of SDS-PAGE (sodium dodecyl sulfate-polyacrylamide gel electrophoresis).

After the proteins have migrated through the gel, they can be visualized using various staining techniques or transferred to a membrane for further analysis. By comparing the migration of the sample proteins to molecular weight standards of known size, the approximate size of the sample proteins can be determined.

Procedure: Protein production and purification using the pET28 system (group exercise)

Materials needed

- To produce the protein of interest
  - A pET28 plasmid containing a 6X histidine-tagged protein gene in the multiple cloning site (many of them are available at addgene.org, e.g., pET 28 full-length human PARP1)
  - *E. coli* BL21
  - Kanamycin 100 mg/ml (stock solution)
  - Isopropyl β-D-1-thiogalactopyranoside (IPTG) 100 mM (stock solution)
  - 10 ml LB broth per group
  - 100 ml LB broth per group
- To run an SDS-PAGE
  - Gel casting apparatus
  - Comb for creating wells

- Spacer plates and clamps
- Electrophoresis chamber and power supply
- 30% or 40% Polyacrylamide/Bis-acrylamide (19:1) solution (e.g., BIO-RAD #1610144)
- Protein size marker
- 2X protein sample loading buffer (Laemmli buffer)
- Gel washing solution (50% (v/v) ethanol, 40% water, 10% (v/v) acetic acid)
- Stain solution (0.1% (w/v) Coomassie blue R350, 20% (v/v) ethanol, and 10% (v/v) acetic acid)
- Storage solution (5% (v/v) acetic acid)

• For protein purification

- Imidazole
- Glycerol solution
- Tris-HCl
- NaCl
- Plastic columns
- Nickel or cobalt-conjugated resin
- Lysis buffer (50 mM Tris-HCl [pH 7.5], 500 mM NaCl, 10 mM imidazole, and 10% glycerol)
- Buffer I (50 mM Tris-HCl [pH 7.5], 500 mM NaCl, 30 mM imidazole)
- Buffer II (50 mM Tris-HCl [pH 7.5], 500 mM NaCl, 50 mM imidazole)
- Elution buffer (50 mM Tris-HCl [pH 7.5], 500 mM NaCl, 250 mM imidazole)
- Glycerol (10%, vol/vol)
- PBS (Phosphate-buffered saline) buffer

• Others

- 10 ml disposable pipettes
- 15 ml conical disposable tubes
- 50 ml conical disposable tubes
- 1.5 ml mini tubes

• Before Day 1

1. The instructor purifies a pET 28 based plasmid (e.g., pET 28 full-length human PARP1) from *E. coli* DHα containing the plasmid and performs

transformation to transfer the plasmid to the *E. coli* BL21 strain. For the transformation, use LB agar plates with kanamycin 50 μg/ml.

2. The instructor picks a single colony on a transformation plate, inoculates in 10 ml LB broth containing 50 μg/ml of kanamycin, and incubate overnight while shaking at 37°C.

Day 1

1. Inoculate 100 ml of LB broth containing 50 μg/ml of kanamycin with 1 ml of overnight culture and incubate while shaking at 37°C until an optical density at 600 nm (OD600) reaches 0.6 (this takes 2 to 3 hours).
   - The concentration of kanamycin stock solution is 100 mg/ml. How much kanamycin stock solution should be added to 100 ml LB broth to make LB broth with 50 μg/ml of kanamycin?
2. Take 10 ml of culture in a 15 ml conical tube.
3. Centrifuge the 10 ml culture at 6000 X g for 10 min, discard the supernatant, and store at -20°C. (This step should be done after step 4.)
4. Add isopropyl β-D-1-thiogalactopyranoside (IPTG) to the ~90 ml culture (final concentration = 0.1 mM) and incubate while shaking for three hours at room temperature.
   - The concentration of the IPTG stock solution is 100 mM. How much IPTG stock solution should be added to the 90 ml culture to make 0.1 mM of IPTG?
5. Take 10 ml of culture in a 15 ml conical tube.
6. Centrifuge the 10 ml culture at 6000 X g for 10 min, discard the supernatant, and store at -20°C.
7. Harvest the remaining 80 ml bacterial culture with centrifugation (6000 X g for 10 min at 4°C) using two 50 ml conical tubes. Remove all the supernatant completely.
8. Store the cell pellets at -20°C.
9. Perform the exercise below during the incubation times.
   a. Download the software SnapGene Viewer to your computer (https://www.snapgene.com/snapgene-viewer).

b. Download the sequence of the plasmid pET 28 full-length human PARP1 from https://www.addgene.org/192635/sequences/ (click the SnapGene link under the plasmid map).

c. Open the sequence with SnapGene Viewer.

d. Analyze the sequence.

- Draw a schematic diagram including the locations of the promoter, operator, ribosome binding site, transcriptional terminator, 6 X Histidine tag location, relative to the PARP1 gene.
- What is the promoter sequence of the PARP1 gene?
- What is the sequence of the lac operator?
- What is the sequence of the ribosome binding site?
- What is the transcription terminator sequence?
- Where is the 6 X Histidine tag attached, the N-terminus or C-terminus of PARP1?
- What is the size (molecular weight) of PARP1?

Day 2

1. Make a 12% (w/v) SDS-PAGE gel.

   Running gel preparation (10 ml running gel solution for 2 minigels)

   Deionized water: 3.3 ml

   30% acrylamide/bisacrylamide: 4.0 ml

   1.5M Tris-HCl, pH 8.8: 2.5 ml

   10% (w/v) SDS: 100 µl

   10% (w/v) APS: 100 µl

   TEMED: 4 µl

Staking gel preparation (5 ml stacking gel solution for 2 minigels)

Deionized water: 3.4 ml

30% acrylamide/bisacrylamide: 0.83 ml

1.0 M Tris-HCl, pH 6.8: 0.63 ml

10% (w/v) SDS: 50 µl

10% (w/v) APS: 50 µl

TEMED: 5 µl

- Assemble the glass plates using the gel caster.
- Prepare the running gel solution as described above. Add APS and TEMED last and mix carefully to avoid the formation of bubbles. APS and TEMED catalyze polymerization.
- Pour the gel solution between the glass plates with a pipette, leaving about ¼ of the space free for the stacking gel.
- Add 100 µl of water on top of the gel and wait until the resolving gel polymerizes.
- Remove the water.
- Prepare the stacking gel solution as described above. Add APS and TEMED last and mix carefully to avoid the formation of bubbles.
- Pour the stacking gel solution carefully with a pipette to avoid the formation of bubbles.
- Insert combs. Allow the gel to polymerize.
- Put the gel into the gel running tank and pour the running buffer into the tank. Make sure that the gel wells are covered with the buffer.
- Remove the comb carefully.

2. You have two bacterial pellets (from cultures before and after adding IPTG). Resuspend the bacterial pellets with 5 ml PBS buffer. Take 1 ml from the sample before the IPTG addition and 0.5 ml from the sample after the IPTG addition and transfer the samples to 1.5 ml mini tubes.
3. Do the centrifugation and remove the supernatant completely.
4. Add 50 ul of 2X protein sample loading buffer (Laemmli buffer).
5. Cap the tubes tightly and heat the sample solutions at 95°C for 10 minutes using a heat block.
6. Centrifuge the samples at 10,000 rpm for 10 minutes.

7. Load 8 μl of the supernatants onto the wells of the 12% (w/v) SDS-PAGE gel.
8. Load 5 μl of the size marker onto a well of the gel.
9. Run the gel at 100–150 V (constant voltage) until the dye front (bromophenol blue) reaches about 1 cm above the bottom of the gel.
10. Pry apart the glass plates, place the gel into a dish on a rocker, and fix the gel with the washing solution for 20 minutes.
11. Rince the gel in water three times, each wash lasting 5 minutes.
12. Incubate the gel in the staining solution for 1 hour.
13. Destain the gel with the washing solution.
14. Once the destaining is completed, store the gel in the storage solution.
15. Take a gel picture.

Day 3

1. Resuspend the cell pellet from 80 ml culture with 10 ml of the lysis buffer.
2. Sonicate the bacteria to lyse (the instructor will show a demonstration).
3. Remove the bacterial debris by centrifugation at 6000 X g for 20 minutes at 4°C.
4. Collect the supernatant in new tubes carefully.
5. Prepare a column loaded with nickel or cobalt resin at the 4°C room.
6. Load the bacterial supernatant onto the column and wait until there is no drip from the column.
7. Wash the column by adding 10 ml of buffer I to the column and wait until there is no drip from the column.
8. Wash the column by adding 10 ml of buffer II to the column and wait until there is no drip from the column.
9. Elute with 1 ml of the elution buffer.
10. Dialyze twice with PBS and once with PBS containing 10% (vol/vol) glycerol.
11. Aliquot the sample and store at -20°C.

Day 4

1. Mix 20 µl of the sample with 20 µl of the 2X protein sample loading buffer in 1.5 ml mini tubes. Cap the tubes tightly.
2. Heat the sample solutions at 95°C for 10 minutes.
3. Centrifuge the samples at 10,000 rpm for 10 minutes.
4. Load 10 µl of sample and 5 µl of size marker onto a 12% (w/v) SDS-PAGE gel.
5. Run the gel at 100–150 V (constant voltage) until the dye front (bromophenol blue) reaches about 1 cm above the bottom of the gel.
6. Pry apart the glass plates, place the gel into a dish on a rocker, and fix the gel with the fixing solution for 20 min.
7. Rince the gel in water three times, each wash lasting 5 minutes.
8. Incubate the gel in the staining solution for 1 hour.
9. Destain the gel with the destaining solution.
10. Once the destaining is completed, take a gel picture.

Review questions

1. List examples of bioproducts produced through genetic engineering (you can do internet search).

2. What is the origin of the RNA polymerase in the pET system?

3. How does IPTG work to promote protein production in the pET system?

4. What is the purpose of the 6X histidine tag attached to the protein expressed in this exercise?

5. What are the differences between the stacking gel and the resolving gel in the polyacrylamide gel, and how do these differences affect protein separation?

6. What is the size of the protein expressed in this exercise? Estimate the size based on the exercise result.

Credits

APPENDIX 1

MICROBIOLOGY LAB SAFETY RULES AND PROCEDURES

Safety in microbiology labs is of utmost importance due to the potential hazards associated with microorganisms. Here are some of the common rules and procedures:

- Do not eat, drink, smoke, chew gum, or apply cosmetics/lip balm in the lab. This prevents ingestion or contact with hazardous materials.
- Place all personal belongings (backpacks, coats, etc.) in the place provided as indicated by the instructor.
- Only wear closed-toe shoes and ensure that lab coats are worn and buttoned.
- Keep hands, fingers, and other objects away from your mouth, eyes, and face. For instance, don't put pens, pencils, or labels in your mouth.
- Do not mouth pipette cultures or reagents.
- Wear gloves when handling cultures or microorganisms. After use, dispose of them in biohazard trash.
- Clean your work area with provided disinfectant before and after each lab session.
- Long hair must be tied back all the time while you are in the laboratory.
- Immediately report injuries to the lab instructor.
- Treat all microorganisms as potential pathogens: always handle cultures with care and do not take them outside the laboratory. If a culture is spilled, immediately notify the lab instructor.
- Properly dispose of contaminated materials.
    - Use biohazard containers for disposing of cultures, gloves, and other contaminated materials. Never discard contaminated materials in regular trash containers.
    - Never pour cultures down the sink.
    - Used glass slides should be placed in the designated containers.

- Before leaving the lab
    - Turn off the gas.
    - Return lab reagents and materials to the original place.
    - Disinfect your lab bench.
    - Wash your hands with soap and water.

These rules ensure a safe working environment and reduce the risks associated with handling microorganisms.

MIX
Papier | Fördert gute Waldnutzung
FSC® C083411

Zeitfracht Medien GmbH
Ferdinand-Jühlke-Straße 7
99095 Erfurt, Deutschland
produktsicherheit@kolibri360.de